Bird life of mountain and upland

Bird Life Series

This series of books will present the bird life of the various habitats in the British Isles in a way that outlines how the birds are adapted to the various environments they inhabit and how the environments themselves shape the birds behaviour and breeding patterns. They thus present a new, ecologically orientated approach to the understanding of the bird life of Britain and Ireland.

Other titles in the series will be

Bird life of freshwater wetlands
Bird life of woodland
Bird life of farmland, grassland and heathland
Bird life of parks and gardens
Bird life of coasts and estuaries

The editor of the series is Dr C.M. Perrins, who is the Director of the Edward Grey Institute of Field Ornithology, Unversity of Oxford.

Bird life of mountain and upland

D.A. RATCLIFFE

With line illustrations by Chris Rose

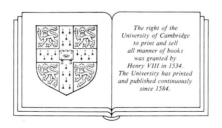

CAMBRIDGE UNIVERSITY PRESS

Cambridge

New York   Port Chester   Melbourne   Sydney

Published by the Press Syndicate of the University of Cambridge
The Pitt Building, Trumpington Street, Cambridge CB2 1RP
40 West 20th Street, New York NY 10011, USA
10 Stamford Road, Oakleigh, Melbourne 3166, Australia

First published 1990

Printed in Great Britain by The Bath Press, Avon

British Library cataloguing in publication data

Ratcliffe, Derek
  Bird life of mountain and upland.
  1. Great Britain. Heathland & mountainous regions. Birds
  I. Title II. Series
  598.2′941

Library of Congress cataloguing in publication data

Ratcliffe, Derek A.
  Bird life of mountain and upland/D.A. Ratcliffe.
    p. cm – (Bird life series)
    Includes bibliographical references.
    ISBN 0 521 38648 9
    1. Birds–British Isles. 2. Birds–British Isles–Ecology.
  3. Bird populations–British Isles. I. Title. II. Series.
  QL690.B65R37 1990
  598.2941–dc20   90-1387   CIP

ISBN 0 521 33123 4 hardback

BO

To Jeannette

CONTENTS

PREFACE

I have always been drawn to the mountains and moorlands above all other types of country. Their birds gain in charm and fascination from the wild and beautiful places in which they live, and seeking them here has a special appeal. When I began, there were few books on the hill birds – mainly accounts by bird photographers of their quest for workable nests and experiences from the hide. I devoured these eagerly and was fired with enthusiasm by their vividness. The regional and county avifaunas were another source of information, though mainly scattered under different species. And there were personal contacts with local ornithologists who had their own extensive and first-hand knowledge of the mountain and moorland birds, as the best source of help and inspiration.

Nowadays there is a wealth of literature and information. Many of the notable species are the subject of whole monographs or long and detailed papers in the scientific journals. Some birds have received enormous attention by many different people, and a great deal is known about them. Almost every year the list of birds hitherto little known grows shorter. There have been numerous general works on the uplands, their ecology and natural history, as well as the birdlife in particular, and these fill in the context. Some species are, nevertheless, still rather little studied and knowledge about them is somewhat general.

This is not a guidebook on where to see upland birds: there are plenty of those already. My background is as an ecologist, and this is the main emphasis I have given to the subject of this book. I have tried to select some of the more interesting and significant aspects of the lives of our mountain and moorland birds, and the ways in which they relate to their environment, including each other. The approach is based on a subdivision of uplands into several main types, with an account of the distinctive though overlapping bird assemblages of these. In the field I have paid attention to several species in particular – Peregrine, Raven, Golden Eagle, Golden Plover and Dotterel – and have written here about some aspects of these from personal knowledge, though drawing on other relevant studies besides. For the rest, my field knowledge varies from casual and patchy to non-existent, and here my treatment has had to be guided by the work and writings of others. I have given most space to those species which may truly be regarded as hill birds, compared with a briefer treatment of others with more debatable claims to this designation.

The authors of this series were told to be sparing with references. This requires advance explanation and apology, in that some major sources for particular species are referenced only once, though clearly some are rather freely drawn upon thereafter. Some significant papers which would merit reference in a more technical work are not mentioned at all, and many points of detail are not acknowledged. Some of the more general major sources, such as *Birds of the Western Palaearctic* (Cramp and Simmons, 1977–88), *Birds of the British Isles* (Bannerman, 1953–63), *Atlas of European Birds* (Voous, 1960), *The Atlas of Breeding Birds in Britain and Ireland* (Sharrock, 1976), *The Atlas of Wintering Birds in Britain and Ireland* (Lack, 1986) and *Birds in Scotland* (Thom, 1986), should be assumed to be a frequent source of factual material. Where a name is given without a date, this refers to a personal communication.

I am greatly in the debt of the many ornithologists and others whose information, ideas and encouragement have helped me so much, and those whose companionship in the field has been an added pleasure. The late Ernest Blezard, naturalist extraordinary, whose fund of knowledge was always a source of wonder, inspired and fostered my early interest. Desmond Nethersole-Thompson has given freely from his vast and unique experience of the mountain birds, to my great benefit. I am grateful to my colleagues in the Nature Conservancy Council for much information and discussion of conservation issues, particularly Des Thompson, John Mitchell, Peter Walters Davies, Peter Davis, Dick Balharry, Derek Langslow, Alison Rush, James Marsden, Mark Felton, Tim Reed, Mike Pienkowski, David Stroud and Jeff Watson. I have also had rewarding discussions of upland birds with Jim Lockie, Ian Newton, Mick Marquiss and Adam Watson. In the Royal Society for the Protection of Birds, Ian Prestt, Art Lance, Roger Lovegrove, Graham Williams, Iolo Williams and Roy Dennis have been extremely helpful in various ways.

My gratitude to numerous providers of Peregrine data has been expressed previously: there is space here for only a collective tribute to most of them, but for much help with other species as well, I thank Geoff Horne, Jim Birkett, Doug Weir, Dick Roxburgh, George Carse, Richard Mearns, Peter Dare, Geoffrey Fryer and Paul Stott. Colin Harrison, Peter Hudson, John Callion and John Strowger have contributed particular information, and I am grateful to the late Bill Robinson, the late Walter Thompson, Stuart Illis, Grant and Jean Roger, and Chris Durell for their help and company in the field. Des Thompson, Donald Watson and John Birks have each read sections of the book, and given valuable advice and suggestions for improvement. The work has also had the benefit of comments from Chris Perrins, as editorial consultant to the series. My wife, Jeannette, has typed the manuscript and given me every encouragement in the writing of it.

I thank the many private landowners and the Forestry Commission for access to the uplands in their ownership. In the concluding chapter on conservation of upland birds, the views expressed are entirely my own, in a private capacity, and are not to be taken as representing those of the Nature Conservancy Council. The politics of conservation are fluid, so that some situations may change appreciably and present commentary upon them soon become out-dated. Such commentary nevertheless seems worth while, as an opportunity for contributing to the public debate which may cause the direction of change to be beneficial.

Chris Rose has most ably met my wishes for illustrations with his wonderful series of drawings and cover painting. I am grateful to Dennis Green, Robert Smith, Edmund Fellowes, Robin Fisher and Eric Hosking for supplying bird photographs. The habitat photographs are my own. For their permission to reproduce maps and figures, I thank the British Trust for Ornithology and Tim Sharrock (breeding distribution in Britain and Ireland), Karel Voous (World breeding distribution) and Peter Hudson (population changes in Red Grouse).

INTRODUCTION

Open and uncultivated hill ground covers between one quarter and one third of Britain and forms by far the largest remaining extent of undeveloped habitat. This expanse of about seven million hectares contains areas which are the nearest condition to real wilderness that our island can show. Our mountains are small-scale and relatively insignificant compared with the great alpine ranges of the World, but they are a much-loved part of the landscape with an endless fascination for mountaineers and naturalists alike. Much of their interest lies in their variety, for no two hill districts are quite alike, and their extremes show striking contrasts. The jagged and precipitous gabbro peaks of the Black Cuillin in Skye could not be more different from the almost flat and tundra-like peat moors of Caithness; and the massive, snowy arctic–alpine tableland of the high Cairngorms contrasts with the peat-clad plateau of central Dartmoor, though both are granite. These differences add diversity to the bird life.

Roughly one quarter of Ireland, covering nearly two million hectares, consists of mountain and upland type habitats, again of widely varying character. Most of the high massifs lie around the periphery of the country, surrounding the large central plain which once contained large and undisturbed peat bogs.

The terms mountain, upland, moorland and hill are all variably and loosely applied to high ground in Britain and Ireland. For the sake of brevity I shall hereafter tend to use *upland* as a general omnibus category to cover all: but the word is not to be understood here in its more limited agricultural usage. Upland habitats are, broadly, the open and unenclosed heaths, grasslands, peat bogs and rocky terrain lying above the limits of cultivation. Land above a certain elevation is implied, but there are difficulties to delineating mountains and uplands by any particular altitudinal limits. The ecological character of the ground, in terms of controlling environment, vegetation and animal community (the ecosystem) is the definitive feature. And there are great variations in the lowest elevations at which distinctively upland ecosystems appear, because they depend on controlling gradients of climate giving marked differences

1

Fig. 1. Map of the distribution of upland habitats in Britain and Ireland (in black); and isotherms of mean annual temperature in ° C, corrected to sea level (re-drawn from the *Climatological Atlas of the British Isles*, London, HMSO, 1952).

across Britain. Mean temperature at sea level at Lands End is 11 °C (52 °F), whereas in north Shetland it is only 7.2 °C (45 °F); and rainfall, which also has a profound effect on agricultural potential and practice, varies from an annual total of 500 mm in East Anglia to over 3000 mm in many western mountain ranges. The uplands contain other important habitats, such as woodlands, lakes and rivers, and in the north and west of Scotland, and western Ireland, they often run continuously down to the sea.

There are few kinds of wildlife whose distribution precisely characterises the uplands. Ling heather, *Calluna vulgaris*, is one of the most distinctive plants, but also covers large areas of heathland in the lowlands of southern and eastern England. The Red Grouse which feeds upon it perhaps best typifies the hill ground, but is often absent from the grassier uplands (Fig. 29). Few plants or animal species confined to hill ground occur in all the British or Irish uplands. Technical definitions can be made according to the detailed species composition of plant and animal communities, but which of these communities to include depends on subjective choice. In general, ground above the limits of enclosed farmland conveniently describes the scope of this book, while remembering that this is an arbitrary definition and that enclaves of similar ground often occur at lower levels within farmed areas. The list of 66 breeding bird species regarded as upland is similarly arbitrary, since some of them are widespread at lower levels and some depend at least partly on other main habitats within upland country, sometimes including farmland (Tables 1 & 2).

One of the characteristic features of mountains, including those of Britain and Ireland, is the tendency to a zoning of habitats with increasing altitude (Fig. 8). Originally, our mountains were mostly covered with a lower zone of forest, except where the ground was especially wet. This forest showed decrease in stature of the component trees in its upper parts, and eventually gave way to scrub and then to heath and grassland as increasing harshness of conditions suppressed the growth of woody plants. The upper forest edge (or tree line) varied in precise altitude, even on different parts of the same mountain, according to differences in shelter and exposure, but even more widely in different parts of the country. Its upper limit may conveniently be regarded as the approximate division between the lower, submontane (or subalpine) and the upper, montane (or alpine) zones. But Man long ago destroyed most of the original forests in these islands, and a true upper tree-line is visible in only a very few places today. Its position usually has to be inferred by the occurrence of scattered trees on cliff faces or by the character of other vegetation.

Many of the lower hill ranges were once wooded right to their summits, and would still be so today, had it not been for Man. The grasslands and heaths which now occupy their place, and are the typical upland habitats, were largely created by human activity. Yet, since they are composed

Table 1. *Miscellaneous data on the breeding birds of the British and Irish uplands*

1	2	3	4	5	6	7	8	9	10
Species	Period on the uplands	Onset of laying	Clutch size	Incubation period (days)	Fledging period (days)	Altitudinal range (m)	Population size (pairs)	Wintering area	Biogeographical type
Black-throated Diver	end March–April to August–early Sept	12 May	2 (1–2)	28–30	60–65	10 – 350	150 decreasing	GB coastal waters, occ. inland	Mid Arctic–Boreal (H)
Red-throated Diver	end March–April to August–early Sept	27 May	2 (1–2)	26–28	38–48	10 – 550	1200–1500	GB & Ireland coastal waters, occ. inland (especially eastern)	High Arctic–Boreal (H)
Mallard	*early March to August–September	18 April	8–12 (5–13)	27–28	50–60	SL – 610	*70 000+	GB & Ireland, lowland inland waters & coasts	Subarctic–Temperate (H)
Teal	early March to August–September	30 April	8–11 (5–13)	21–23	25–30	SL – 730	*3500–6000	GB & Ireland, lowland inland waters & coasts	Low Arctic–Temperate (H)
Wigeon	end March–April to August–September	7 May	8–10 (6–12)	24–26	40–45	10 – 600	300–500	GB & Ireland, lowland inland waters & estuaries	Low Arctic–Boreal (P)
Common Scoter	late April–May to early August	early June	6–8 (5–11)	27–31	45–50	50 – 230	160–190	GB & Ireland coastal waters	Low Arctic–Boreal (H)
Red-breasted Merganser	March–April to early June (♂) August–Sept (♀)	25 May	8–10 (6–14)	31–32	60–65	SL – 350	2000–3000	GB & Ireland coastal waters, occ. inland	Low Arctic–Boreal (H)
Goosander	end February–March to late May (♂) August–Sept (♀)	25 April	8–11 (4–13)	30–32	60–70	10 – 600	900–1300 increasing	GB lowland inland waters, occ. coastal areas	Subarctic–Boreal (H)
Greylag Goose	All year	18 April	4–6 (3–12)	27–28	50–60	SL – 300	*600–1000	GB & Ireland, lowland inland habitats and estuaries	Subarctic–Temperate (P) disjunct
Golden Eagle	All year	1 April	2 (1–3)	43–45	65–75	SL – 915	430 decreasing	Breeding areas, juveniles move into uplands farther afield	Boreal–Subtropical (H)
Buzzard	All year	25 April	3 (2–5)	32–38	50–55	SL – 600	*12 000–15 000	Breeding areas, juveniles move into other areas	Boreal–Tropical (H & S)
Red Kite	All year	7 April	2 (1–3)	31–32	50–60	100 – 400	40–50 increasing	Breeding areas, juveniles move into other areas	Temperate–Mediterranean (P)
Hen Harrier	*mid-March to mid-August	5 May	4–6 (3–7)	29–31	32–42	10 – 520	400–600 decreasing	GB & Ireland, lower hill ground, marginal land, lowland coastal areas	Subarctic–Temperate (H & S)

Species	Breeding season	First egg date	Clutch	29–32	38–46	Altitude (m)	Population	Distribution	Zone
Peregrine	All year	8 April	3–4 (2–5)	29–32	38–46	SL – 975	*1000+ (c.650 upland) increasing	Breeding areas, some pairs and juveniles move to lower or more southerly ground	Low Arctic–Tropical (C)
Merlin	*March to August –September	10 May	4–5 (3–6)	26–28	26–32	SL – 600	550–650 decreasing	more southerly ground GB & Ireland lowlands & coasts, occ. movement to mainland Europe	Subarctic–Boreal (H)
Kestrel	All year	1 May	4–6 (3–7)	27–29	27–32	SL – 600	*50 000–80 000	Lower hill ground, & juveniles move widely into lowlands	Boreal–Tropical (OW)
Red Grouse	All year	20 April	6–10 (4–12)	20–24	30–35	SL – 820	300 000 decreasing	Breeding areas, strongly sedentary	Low Arctic–Boreal (H)
Ptarmigan	All year	15 May	5–8 (3–10)	21–23	28–33	350 – 1265	10 000	Breeding areas, descending lower in severe weather	High Arctic–Alpine (H)
Black Grouse	All year	28 April	6–11 (4–12)	25–27	35–40	SL – 500	10 000–30 000 decreasing	Breeding areas, strongly sedentary	Boreal–Temperate (P)
Oystercatcher	early March to end July	1 May	2–3 (1–4)	24–27	28–32	SL – 490	*40 000–50 000	Coastal GB, Ireland, France, Iberia	Subarctic–Temperate (C) disjunct
Lapwing	mid-February to June–early July	7 April	4 (3–5)	26–29	35–40	SL – 610	*215 000 decreasing	GB, Ireland, France, Iberia	Boreal–Temperate (P)
Golden Plover	*mid February to early July	20 April	4 (3–4)	28–32	25–33	10 – 1040	28 000	Mainly GB & Ireland, fewer in W Europe	Low Arctic–Boreal (P)
Ringed Plover	end March to late July–August	12 May	4 (3–4)	23–25	23–26	SL – 430	decreasing *8 600	Coastal GB, Ireland, France, Iberia	High Arctic–Temperate (H)
Dotterel	late April to mid August	27 May	3 (2–3)	24–28	25–30	600–1200	500+ increasing	N. Africa & Middle East	Mid Arctic–Alpine (P) disjunct
Snipe	late March to end August	1 May	4 (3–4)	18–20	19–20	SL – 800	*40 000 decreasing	GB, Ireland, NW Europe	Low Arctic–Tropical (H & S) disjunct
Curlew	February–March to late June–July	27 April	4 (3–4)	28–30	32–40	SL – 650	*45 000–50 000	Western GB & Ireland, some to coast of France & Iberia	Boreal–Temperate (P)
Whimbrel	late April–early May to late July	20 May	4 (3–4)	27–28	30–38	SL – 300	500 increasing	Coast of Africa, especially the west	Low Arctic–Boreal (H)
Wood Sandpiper	early May to July	late May	4	22–23	c. 30	120 – 560	3–12	Africa, S of Sahara	Low Arctic–Boreal (P)
Common Sandpiper	mid-April to August	12 May	4 (3–4)	21–22	26–28	SL – 930	22 000–25 000	Africa, S of Sahara	Subarctic–Temperate (P)
Redshank	mid-March to end June–July	end April	4 (3–4)	23–24	25–35	SL – 730	*34 000–37 000	Mainly coastal GB & Ireland, some NW Europe	Subarctic–Temperate (P)
Greenshank	end March–early April to late June–July	4 May	4 (3–4)	23–26	25–31	30 – 685	1545 decreasing	Ireland & western GB, Africa, Middle East	Subarctic–Boreal (P)
Dunlin	early April to late June–July	4 May	4 (3–4)	21–23	19–21	SL – 1070	9000– 10 000 decreasing	North-west and Tropical West Africa	Mid Arctic–Boreal (H)

Table 1. (*continued*)

Species	Period on the uplands	Onset of laying	Clutch size	Incubation period (days)	Fledging period (days)	Altitudinal range (m)	Population size (pairs)	Wintering area	Biogeographical type
Purple Sandpiper	early May to end July–August	early June	4 (3–4)	21–22	c. 21	800–1200	1–4	Coastal GB, Ireland & NW Europe	High Arctic–Subarctic (H)
Temminck's Stint	mid-late May to August	mid June	4 (3–4)	19–22	17–18	100–320	<6	Mediterranean, Middle East & Tropical Africa	Low Arctic–Subarctic (P)
Ruff	end April–May to July	end May	4 (3–4)	20–24	25–28	120?	*10–12	W coastal Europe & Africa	Low Arctic–Temperate (P)
Red-necked Phalarope	mid-May to late August	early June	4	17–21	20	SL–125	19 decreasing	Arabian Sea	Low Arctic–Subarctic (H)
Arctic Skua	late April to mid August	25 May	1–2	25–28	25–30	10–380	3400 increasing	Open sea: North & South Atlantic	High Arctic–Subarctic (H)
Great Skua	early April to mid August–September	18 May	1–2	27–31	40–51	10–380	7900 increasing	Open sea: North & South Atlantic	Antarctic–N Atlantic disjunct
Great Black-backed Gull	end March to late July	5 May	3	27–29	42–56	SL–500	*25 000	Local movement S within GB & Ireland, a few to France & Iberia	Northern Atlantic (H)
Lesser Black-backed Gull	late March to July–August	1 May	3	25–27	35–40	SL–560	*85 000 increasing	Coastal France, Iberia, W Mediterranean, NW Africa	Low Arctic–West European (P)
Herring Gull	early March to end July–August	1 May	3	25–27	35–40	SL–560	*<300 000 decreasing	Local movement S, mostly remaining in GB & Ireland	Low Arctic–Mediterranean (H) disjunct
Common Gull	March–April to July	10 May	3	23–25	35	SL–870	*50 000	Local movement S & SW, tending to reach Ireland & Irish Sea	Low Arctic–Boreal (H)
Black-headed Gull	mid March to July	1 May	3	22–24	35	SL–870	*150–300 000	Wide but variable movements within GB & Ireland, a few to Europe	Subarctic–Temperate (P)
Stock Dove	mid March to end July–early August	20 April	2	16–18	20–30	10–460	*100 000	Local movements to adjoining lowlands in GB & Ireland	Boreal–Temperate (P)
Cuckoo	late April to July–August	20 May	9–12 ×1	12–13	19–23	SL–600	*17 500–35 000	Central & South Africa, Saudi Arabia	Boreal–Tropical (OW)
Short-eared Owl	*early March to August	25 April	4–7 (3–14)	25–27	24–27	SL–550	*1000	Movement to lowlands & coasts of GB, and some to adjoining Europe	Low Arctic–Temperate (H & S)

Species								Movement around Shetland Isles	High–Low Arctic (H)
Snowy Owl	All year	mid May–early June	5 (1-7)	30-33	43-50	150	0-1		High–Low Arctic (H)
Nightjar	late May to August	mid June	2	17-18	16-18	10 – 370	*<2100 decrease since 1940	Africa	Boreal–Mediterranean (P)
Skylark	late February to September	7 May	4 (3-5)	11-12	18-20	SL – 900	*2 000 000+	Local movements to adjoining lowlands in GB & Ireland	Boreal–Mediterranean (P)
Shore Lark	end April to September	end May	2-4	10-11	16-18	1000?	0-1	East coast of Britain, NE and central Europe	Mid Arctic–Temperate (H) disjunct
Raven	All year	5 March	4-6 (3-7)	20-21	35-42	SL – 750	*4000-5000 decreasing	Breeding areas, sedentary, juveniles disperse to other areas	Mid Arctic–Tropical (H)
Carrion & Hooded Crow	All year	20 April	3-5 (3-6)	18-21	30-35	SL – 550	*1 000 000	Breeding areas, sedentary, juveniles disperse to other areas	Subarctic–Temperate (P)
Jackdaw	All year	25 April	4-6	17-18	30-35	SL – 370	*500 000+	Winter flocks in lowlands adjoining upland breeding areas	Boreal–Mediterranean (P)
Chough	All year	28 April	4-5 (3-6)	17-18	30-35	SL – 450	*900-1000	Sedentary, local movements adjoining breeding areas	Alpine–Mediterranean (P)
Wren	All year	7 May	4-6 (3-7)	14-15	16-17	10 – 900	*3- 5 000 000	Some movement to lower ground within or beyond the uplands	Subarctic–Temperate (H)
Dipper	All year	25 March	4-5 (3-6)	15-17	20-24	10 – 750	*c.30 000 local decrease	Breeding areas, strongly sedentary	Boreal–Temperate (P) disjunct
Ring Ouzel	late March to late September	18 April	4 (3-5)	12-15	14-16	50 – 915	8000– 16 000 ?decreasing	Mediterranean – NW Africa	Subarctic–Alpine (P)
Wheatear	late March to end August	12 May	4-7 (2-8)	12-14	14-16	SL – 1220	*80 000 decreasing	Tropical Africa	Mid Arctic–Temperate (H)
Stonechat	All year, or late March to August	1 April	4-6 (2-7)	13-14	12-15	SL – 500	*30- 60 000	From breeding areas to southern & coastal Britain, and W Europe	Boreal–Tropical (OW) disjunct
Whinchat	late April to early September	12 May	4-7	12-14	12-14	10 – 450	*20- 40 000	Tropical Africa	Boreal–Temperate (P)
Meadow Pipit	late March to mid July–early September	7 May	4-5 (3-5)	12-14	13	SL – 1220	*3 000 000	General movement to lowlands & southwards in GB & Ireland	Subarctic–Temperate (P)
Pied Wagtail	end February to late September	20 April	5-6 (3-7)	13-14	15-16	10 – 600	*500 000	Southwards movement to lowlands GB & Ireland, France, Iberia, N Africa	Low Arctic–Tropical (P)

Table 1. (*continued*)

1 Species	2 Period on the uplands	3 Onset of laying	4 Clutch size	5 Incubation period (days)	6 Fledging period (days)	7 Altitudinal range (m)	8 Population size (pairs)	9 Wintering area	10 Biogeographical type
Grey Wagtail	mid March to early September	20 April	4-6 (3-7)	13-14	15-16	10 – 670	*25–50 000	Southwards movement to lowlands GB & Ireland, France, Iberia, N Africa	Boreal–Temperate (P)
Twite	end April–May to August–September	late May	5-6 (4-7)	12-13	15	SL – 500	20–40 000	Moves to coasts, of Scotland, NW & E England, W Ireland	Boreal–Temperate (P) disjunct
Lapland Bunting	end April to August	end May	5 (4-6)	13-14	12-15	c.915	0-16	Coastal areas of GB & Ireland especially in the east	High Arctic–Subarctic (H)
Snow Bunting	All year?	4 June	5-6 (4-7)	12-15	12-14	800-1200	6-80 increased since 1960	Scottish breeders may remain in the uplands	High Arctic–Subarctic (H)

1. *Species*

2. *Period on the uplands*
   Differences in weather can cause this to vary from year to year. Movements may also vary between regions. Those species marked * sometimes winter on lower moors or around the edges of upland massifs, and some of those which are generally sedentary show some degree of movement to lower ground, especially during hard winters.

3. *Onset of laying*
   This aims to give an average date by which at least 10% of the total breeding population has begun to lay. There can be variations between years according to weather, between regions for widespread species, and according to altitude.

4. *Clutch size*
   Normal; range in brackets

5. *Incubation period* and

6. *Fledging period*
   Variation and duration tend to increase with size of species, and according to external factors, especially weather, food supply and disturbance.

7. *Altitudinal range*
   This is based on the normal range for regular breeding and omits some isolated extreme records. Those species breeding on sea cliffs are given as sea level (SL) though in practice they nest above the zone of heavy wave action.

8. *Population size*
   The most recent figures available are given. Those species marked * have a distribution substantially wider than the uplands, but the size of their upland populations is not known. Includes Ireland.

9. *Wintering area*
   As far as possible, this indicates the movement of the population in Britain and Ireland, but for some species only the generalised winter distribution is known for a much larger European population.

10. *Biogeographical type*
    This refers to World breeding distribution and is assessed from sources named on p. 00.

    H, Holarctic
    P, Palaearctic
    OW, Widespread in the Old World
    S, Southern Hemisphere
    C, Cosmopolitan

Table 2. *Distribution of upland birds within Britain*

	Shetland	Orkney	Outer Hebrides	Sutherland & Caithness	Ross & Cromarty	West Inverness & Skye	East Grampians & Cairngorms	West Grampians	Argyll (inc Bute – south & mid Ebudes)	West Southern Uplands	East Southern Uplands	Cheviots	Lake District	Northern Pennines	Mid-Pennines & Bowland Fells	North York Moors	Peak District	North Wales	South & Mid Wales	Dartmoor & Exmoor
Black-throated Diver	+	+	+	+	+	+	o	+	+	o										+
Red-throated Diver	+	+	+	+	+	+	o	+	+	o										o
Mallard	+	+	+	+	+	+	+	+	+	+	+	+	+	+	+	+	+	+	+	
Teal	o	o	+	+	+	+	+	+	o	+	+	+	+	+	+	+	+	+	+	
Wigeon	+	o	+	+	o	+		+	+	o	+	o		+						
Common Scoter	+	+	+	+	+	+		+	+		+						?		?	
Red-breasted Merganser				+	+	+	+	+	+	+	+	+	+	+	+			+		+
Goosander			+	+	+	+	+	+		+			+		+			+		o
Greylag Goose			+	+	+	+	+	+	+	+	o	o	+			o				
Golden Eagle			+	+	+	+	+	+	+	+	+	o	o	+				+	+	
Buzzard		o*		+	+	+	+	+	+	+	+	+	+	+	+	?	?	o	o	+
Red Kite		o*																		+
Hen Harrier	+*	+*	+*	+	+	o	+	+	+	+	+	+	?	+	+	+	+	+	+	o
Peregrine	+	+	+	+	+	+	+	+	+	+	+	+	+	+	+	+	+	+	+	
Merlin		+	+	+	+	+	+	+	+	+	+	+	+	+	+	+	+	+	+	+
Kestrel	+	+	+	+	+	+	+	+	+	+	+	+	+	+	+		+	+	+	+
Red Grouse		+	+	+	+	+	+	+	+	+	+	+	+	+	+	?	+	+	+	
Ptarmigan				+		o	+	+		+	+	+								
Black Grouse				+	+	+	+	+	+	+	+	+	+	+	+	+	?	+	o	o
Oystercatcher	+	+	+	+	+	+	+	+	+	+	+	+	+	+	+			+	+	+
Lapwing	+	+	+	+	+	+	+	+	+	+	+	+	+	+	+	+	+	+	+	+
Ringed Plover	+	+	+	+	+	+	+	+	+	+	+	+		+						
Golden Plover	+	+	+	+	+	o	+	+	+	+	+	+	+	+	+	+	+	+	+	+
Dotterel				+		+	+	+	o	o	o		o	o	o			o		
Snipe	+	+	+	+	+	+	+	+	+	+	+	+	+	+	+	+	+	+	+	+
Curlew	+	+	+	+	+	+	+	+	+	+	+	+	+	+	+	+	+	+	+	+
Whimbrel	+	o	o	?	o		+	+												

Table 2. (continued)

	Shetland	Orkney	Outer Hebrides	Sutherland & Caithness	Ross & Cromarty	West Inverness & Skye	East Grampians & Cairngorms	West Grampians	Argyll (inc Bute – south & mid Ebudes)	West Southern Uplands	East Southern Uplands	Cheviots	Lake District	Northern Pennines	Mid-Pennines & Bowland Fells	North York Moors	Peak District	North Wales	South & Mid Wales	Dartmoor & Exmoor
Wood Sandpiper	+	+	+	+	+	+	+	?	0											0
Common Sandpiper	+	+	+	+	+	+	+	+	+	+	+	+	+	+	+	+	+	+	+	+
Redshank	0	0	+	+	+	+	+	+	+	+	+	+	+	+	+	+	+	+	+	+
Greenshank			+	+	+	+	+	+	+									+	+	
Temminck's Stint				0	0		0	0												
Purple Sandpiper				?			+													
Dunlin	+	+	+	+	+	+	+	+	+	+	+	+		+	+		+	+	+	+
Ruff		0	?	0	+				0											
Red-necked Phalarope	+	0	+	0					0											
Great Skua	+	+	+	?																
Arctic Skua	+	+	+	+					+			?								
Great Black-backed Gull	*+	*+	+	+					+				0		+				?	
Lesser Black-backed Gull	*+	*+	+	+			+	+	+	+	+	?		+	+			+	+	+
Herring Gull	+	+	+	+			+	+	+	+	+	+	+	+	0		+	+	+	+
Common Gull	+	+	+	+			+	+	0	+	+	+	+	+	+	+	+	+	+	
Black-headed Gull	+	+	+	+			+	+	+	+	+	+	+	+	+	+	+	+	+	
Stock Dove							+	+		+	+	+	+	+	+	+	+	+	+	+
Cuckoo	0	+	+	+	+	+	+	+	+	+	+	+	+	+	+	+	+	+	+	+
Snowy Owl																				
Short-eared Owl		+	+	+	+	+	+	+	+	+	+	+	+	+	+	+	+	+	+	+
Nightjar		+					?	?	+				?		?		?		+	+
Skylark	+	+	+	+	+	+	?	+	+	+	+	+	+	+	+	+	+	+	+	+
Shore Lark							0		0						0		0			
Raven		+	*+	+	+	+	+	+	+	+	+	+	+	+	+	+	+	+	+	+
Carrion & Hooded Crow	*+	*+	+	+	+	+	+	+	+	+	+	+	+	+	+	+	+	+	+	+
Jackdaw	+	*+	+	+	+	+	+	+	+	+	+	+	+	+	+	+	+	+	+	+
Chough									*+									+	0	

Species																				
Wren	+	+	+	+	+	+	+	+	+	+	+	+	+	+	+	+	+	+	+	+
Dipper	+	+	+	+	+	+	+	+	+	+	+	+	+	+	+	+	+	+	+	0
Ring Ouzel	+	+	+	+	+	+	+	+	+	+	+	+	+	+	+	+	+	+	0	0
Wheatear	+	+	+	+	+	+	+	+	+	+	+	+	+	+	+	+	+	+	+	+
Stonechat	+	+	+	+	?	?	+	+	+	+	+	+	+	+	+	+	+	+	+	
Whinchat	+	+	+	+	+	+	+	+	+	+	+	+	+	+	+	+	+	+		
Meadow Pipit	+	+	+	+	+	+	+	+	+	+	+	+	+	+	+	+	+	+	+	+
Pied Wagtail	+	+	+	+	+	+	+	+	+	+	+	+	+	+	+	+	+	+	+	+
Grey Wagtail	+	+	+	+	+	+	+	+	+	+	+	+	+	+	+	+	+	+	0	0
Twite		?	+	+	+	+	+	?	?	+	+	+	+	+	+	+	+	+	+	+
Snow Bunting													0	+	0	0	0		0	
Lapland Bunting														0						
TOTAL (66 species)	26	34	39	29	27	38	39	35	35	39	43	47	50	51	44	48	51	45	38	34
	(4)	(2)	(2)	(1)	(1)	(3)	(2)	(3)	(3)	(2)	(4)	(4)	(2)	(4)	(3)	(4)	(4)	(2)	(9)	(4)

+ Well-established and regular breeder

? Uncertain breeder at present

Records are from BTO Atlas (1968–72) onwards: older records and extinctions omitted or given as '?'.

North Wales: north of Dovey estuary

Mid-Pennines: between M62 and Wensleydale–Dent

Northern Pennines: includes Howgill Fells and Shap Fells east of M6

East Southern Uplands: east of Annandale

Argyll: south of L.Linnhe and including Islay, Jura, Mull, Tiree & Coll

West Grampians: Dunbarton, Stirling, Perth W of A9, Inverness W of A9 and south of Great Glen

East Grampians: east of A9

West Inverness: includes mainland Argyll north of L. Linnhe and the Small Isles

0 Occasional breeder or only 1 or 2 pairs

* Coastal cliff breeder within this region

largely of native species and resemble closely the communities associated with open hill woodland, we call them semi-natural. At the foot of the hills, much of the ground cleared of forest was enclosed and became the improved pastureland and hay meadows of the hill farms. The fields tended to be delineated by dry stone walls rather than hedges, and their use was for stock-rearing, rather than arable cultivation. The farmlands and the unenclosed hill are not always clearly separated ecologically. There is in many areas an intermediate type of transition ground, usually encompassed by fence or wall, but often in extensive units, and consisting of poor, unimproved and often coarse or rush-infested pasture. This is the marginal land of the moorland edge and the *ffridd* of the Welsh hills. In many places, it has been the frontier of human advance, pushing forward at times to reclaim more of the open hill, and then retreating again as the struggle to maintain this step became too much. Often there are the ruins of dwelling places and enclosures to tell the story of Nature's readvance.

The bird life of the uplands has a history which has become closely interwoven with human affairs, for so much of the ground above the farms has been used over the centuries as grazing land. The domesticated stock ranging over the hills has variously been of goats, cattle, sheep and ponies. Yet in some areas, the native animals have also been exploited as a crop, though with interest especially in the sport of hunting them. Large areas of the uplands have been managed as preserves for Red Grouse or as grazing range for Red Deer. Almost universally, fire has been a tool of management of the hill grazings, to an extent that has often produced a kind of vegetation with the greatest resistance to further burning. Only the highest Highland tops have escaped this particular kind of heavy exploitation, for their vegetation is too sparse and lacking in nutritive value for it to be greatly sought after by the grazing animals. This was once a terrain of little concern to anyone, but in recent years it, too, has become subject to development.

There are, as we shall see, many different pressures. For we are a population of 56 or so million people in an island of 23 million hectares or, in old-fashioned measure, about one acre per person on average. It is, accordingly, hardly surprising that there are few places on our land surface, however remote, where someone has not worked out that the ground could be put to better use than at present. Change or intensification in land use always causes change for wildlife, including the birds. Sometimes there is a replacement of existing species by others, but often there is simply a decline or a disappearance leading to impoverishment of the previous bird community. The widespread reclamation of moorland to farmland and its even more extensive afforestation are changes with profound consequences for birds. They are at the heart of our modern concern for the conservation of nature; and the needs and measures for ensuring that as much as possible of our mountain bird fauna survives into the future are a vital issue. Human impact is all-pervading, and affects even the remotest regions in some form: there is growing evidence that even climate is subject to its influence.

Interest in the birds of the hill country has been part of the general growth in enthusiasm for field ornithology. Our upland bird fauna is less rich than that of the mountains and tundra of northern Europe, but some of its most distinctive members are of outstanding interest. Among the predators, the waders and the grouse, some species reach population densities which are remarkable compared with most other regions of the World. The total populations of Peregrines and Golden Eagles are among the largest in Europe. We do not have any endemic birds, but the Red Grouse and the Golden Plover have gone some way to evolving insular races adapted to our southern conditions. There is much ecological interest in the ways in which some species have taken advantage of the man-made changes so extensively wrought on the upland scene down the centuries. The persistence of relict and fringe populations, including migrants, of birds which belong to altogether colder northern regions, is a fascinating subject in itself. The remnant populations, in their characteristic habitats, are reminders that this country still has outliers of the Boreal and Arctic environment which occupied the whole of Britain and Ireland at the end of the last Ice Age. The history of their changes, in response to improving climate, is only imperfectly known, but we can at least now watch and understand the readjustments to changing conditions that continue to occur.

Our knowledge of upland birds grew slowly from the time – roughly around the end of the eighteenth century – when people began to be seriously interested in ornithology and to record their observations. For a long time, little particular interest was taken in this group as such, and information about hill birds increased as part of the general growth of ornithological knowledge. The early Victorians began to build the foundations. The tireless walker, William Macgillivray of Aberdeen, explored the mountains and compiled an early systematic treatise, while the gentleman collectors such as Charles St. John, John Wolley and Edward Booth conducted grand tours and wrote vividly of their experiences. The period of county and regional avifaunas followed, giving more particular information on distribution and general abundance in different parts of Britain. The doyen of this treatment was John Harvie-Brown, a man ahead of his age, whose inspiration and industry in compiling, or stimulating others to compile, the fine series of regional faunas of Scotland, is one of the great milestones in Britain's natural history. In northern England, the Rev. H.A. Macpherson produced his scholarly *Vertebrate Fauna of Lakeland* (1892), while the hunter naturalist Abel Chapman wrote perceptively on the birds of the Border moorlands (1907, 1924). In Wales George Bolam (1913) set down many original observations on the hill birds of Merioneth, and the *Birds of Ireland* by Richard Ussher and Robert Warren (1900) became the chief source of information for that country. The Red Grouse Inquiry, set up in 1905, and its great Report of 1911, was a landmark, as probably the first ecological study of a bird species, and the forerunner of the modern scientific approach.

The period 1880–1940 was the great age of egg collecting. Ornithological energies were poured into searching for birds and their nests, especially the rarer kinds and those laying variably marked eggs. Some of the hill species became an especial challenge, and a good deal was discovered and recorded by the 'eggers' about distribution, breeding biology and behaviour. John Walpole-Bond, Cecil Stoney, Norman Gilroy, Francis Jourdain and H.A. Gilbert were amongst those who expanded knowledge of the hill birds. This was also the period during which bird photography developed, and many interesting insights were set down in the growing number of illustrated books on birds in their native haunts, beginning with pioneers such as Richard and Cherry Kearton, Oliver Pike, Arthur Brook, Ralph Chislett and George Kearey. One of the most prolific photographer–authors, and the leading spirit in mountain ornithology, was Seton Gordon, who wrote evocatively about his beloved Cairngorms and other Highland ranges, and inspired others to follow.

The 1930s saw the blossoming of scientific ornithology and since 1945 there has been an enormous increase in attention to hill birds, with in-depth studies of many species now reported. A monograph of the Greenshank by Desmond Nethersole-Thompson in 1951 was followed by his other works on Snow Bunting and Dotterel and Highland birds in general, and new books on the Greenshank and British waders written jointly with his wife Maimie; besides his contributions on the breeding biology and behaviour of many other upland species to the *Handbook of British Birds* (Witherby *et al.*, 1938) and *The Birds of the British Isles* (Bannerman, 1953–63). Also in the Highlands, Adam Watson has undertaken life-long research on Ptarmigan, Red Grouse, Golden Eagle and other species. These two veteran ornithologists joined forces in writing *The Cairngorms* (1981), an account of wildlife in this key mountain region, dealing comprehensively with its birds.

Studies of hill birds have multiplied, from the work of dedicated loners to the collective and long-term efforts of teams. The Red Grouse has received particular attention and there is now a considerable literature on this bird alone. The universities and other institutions of learning, and the government agencies such as the Nature Conservancy Council (NCC) and Institute of Terrestrial Ecology, have played an important part in research developments. The tremendous growth in enthusiasm for ornithology as a spare-time interest has been reflected in the spectacular rise in membership of the Royal Society for the Protection of Birds (RSPB) from 6000 in 1964 to half a million in 1988. This interest has been canalised into survey, study of movements (through ringing) and other research, especially through the British Trust for Ornithology (BTO) and Wildfowl Trust (WT). It becomes increasingly invidious to single out names, for so many have contributed to the recent burgeoning of knowledge about our hill birds. This book attempts to summarise some of the accumulated information, and will refer to many particular contributions. It tries to convey something of the interest of this bird fauna, with the varied influences which rule the lives of the different species and cause their distribution and status to change with time. This in turn leads into our concern for the future of the birds in these areas, regarded by many as Nature's last stronghold in Britain and Ireland; and to the need to conserve what remains of this fascinating part of our wildlife heritage.

1

The birds and their habitat

The upland bird fauna

There is no definitive list of mountain and upland birds, so that my choice of species is open to argument. The designation implies an association with the open ground conditions typical of higher altitudes, but where is the limit to be drawn downhill in relation to other habitats such as woodland? Prime candidates are the strictly montane species, living mostly above the potential tree-line, but of the regular breeders during recorded history (mainly after AD 1800) only the Ptarmigan, Dotterel and Snow Bunting qualify. We have recently been able to add as rare or sporadic montane breeders the Snowy Owl, Purple Sandpiper, Shorelark and Lapland Bunting.

Several other birds of largely treeless tundra in northern Europe belong mainly to submontane moorland in Britain and Ireland: the widespread Red Grouse (representing a southern, insular race of the Willow Grouse), Golden Plover and Dunlin; the more local Twite; and the much rarer or highly localised Whimbrel, Red-necked Phalarope, Arctic Skua and Great Skua. Still other species are typical of partly wooded bog and tundra in the Boreal–Subarctic regions, albeit mostly within the potential forest zone. The Greenshank is widespread in parts of the Highlands, but the Wood Sandpiper is rare and Temminck's Stint perhaps not a regular breeder; a trio found in the open forest heaths and bogs of Scandinavia. The Ruff has very recently been added to the list of Scottish peatland breeders and also belongs especially to the northern European forest marshes. The more widespread Ring Ouzel, Golden Eagle and Merlin have open forest habitats in Scotland, but their main haunts are in unwooded country and the first two extend into the montane zone in some places.

Nearly all these birds of submontane habitats breed down to low elevations, often in coastal habitats and close to sea level, in northern Scotland. Some are also at low levels in western Ireland. The Dunlin is abundant on the coastal marshes of the Outer Hebrides, while most of our Whimbrel, Red-necked Phalaropes and the two Skuas nest on the maritime grass-

16

lands, heaths and marshes of our far northern and western islands. Phalaropes also breed near sea level in the far west of Ireland. The Chough is a hill bird only in Wales, the Isle of Man and a few parts of western Ireland, and is otherwise restricted to rocky coasts with adjoining heath and grassland. Twites are now most numerous in a range of coastal habitats, rather than on inland moors. While the Hen Harrier occupies a wider range of lowland habitats on the European continent, it is mainly an upland nester in Britain and Ireland. The same is true of the Short-eared Owl, except that it is strangely absent from Ireland as a breeder.

There is also a fairly large group of more opportunistic species which occur in a variety of lowland habitats, but find major niches in the mountains and moorlands. The exciting group of predators nesting on the upland cliffs – Peregrine, Raven, Buzzard and Kestrel – are equally at home on coastal precipices or inland rocks in low-lying situations, provided there is suitable feeding habitat nearby. The last three are equally adapted to tree-nesting in both lowland and upland environments, and even the Peregrine is known to nest in trees. Carrion or Hooded Crows are widespread, and other characteristic hill-going passerines are the Cuckoo, Meadow Pipit, Skylark, Wren, Wheatear, Whinchat and Stonechat. The lower hills and marginal land have become increasingly important for Lapwing, Snipe and Redshank as the lowland farm habitats have been 'improved', while the Curlew has moved downwards from the uplands to become equally a bird of the lowlands. Black Grouse and Red Kite also have strong claims to be considered mainly hill birds nowadays.

A further group of birds is associated with lakes and streams in upland and/or northern areas. Besides some of the waders already mentioned, it includes Wigeon, Goosander, Red-breasted Merganser, Common Scoter, Red-throated Diver, Black-throated Diver, Greylag Goose, Common Gull, Common Sandpiper, Dipper and Grey Wagtail. The Mallard, Teal and Black-headed Gull are also as much upland as lowland waterfowl.

Beyond this, it becomes very much a matter of opinion over which other species might justifiably be added. I feel that there is a case for giving at least brief mention to the following, as having a significant or especially interesting foothold in the uplands: Stock Dove, Jackdaw, Pied Wagtail, Nightjar, Oystercatcher, Ringed Plover, Herring Gull, Lesser Black-backed Gull and Great Black-backed Gull.

I have drawn the line here, omitting species of northern forest, such as the Fieldfare, Redwing, Capercaillie, and also a few of lakes and rivers within the forests or along the main valley floors, such as the Osprey, Slavonian Grebe, Goldeneye and Whooper Swan. Bluethroats have nested recently but I do not know the habitat. I have not included birds which, although characteristic of lower hillslopes in some areas, are almost invariably associated with at least open growths of trees, such as the

Willow Warbler, Tree Pipit, Redstart, Chaffinch and Tawny Owl. And although the Heron locally follows hill streams and lakes, it is essentially a woodland and lowland nester. The Grey Partridge, Yellowhammer, Reed Bunting and Grasshopper Warbler occasionally nest up to moderate elevations on open hill ground in more southerly upland districts, but are not constant enough to merit inclusion. There is, however, a discussion of the occasional or local adaptation of other birds to upland breeding niches, as the possible pioneers of ecological divergence into this class of habitats.

The bird fauna of the uplands is the product of interactions between different conditions of climate, geology and topography; the modifying effect of man's activities; and the adaptations evolved by the birds themselves over past ages. While it reflects a present pattern of environment, it cannot be adequately understood other than as a long historical process of response to change, especially of climate and growing human impact. This chapter will examine some of these ecological and historical factors, and their connections.

It could hardly be claimed that these are particularly rich habitats for birds. Of at least 230 species which have bred in Britain and Ireland since 1967, only 66 are regarded as candidates for my list. The rapid altitudinal deterioration of a generally harsh climate gives a rather hostile higher level environment which few forms can tolerate. Only 17 species breed regularly or in any numbers above the potential tree-line (Chapter 7) and only the Ptarmigan lives mainly within this montane zone all the year round. The lower, submontane moorlands have many more species, as well as a much larger biomass per unit area, but they vary a good deal in these respects.

As in other habitats, the effect of conditions on food supply is overridingly the key to the character, distribution and abundance of upland birds. The quality of environment as nesting habitat is also of major importance.

The importance of climate

Climate is of fundamental importance in setting the limits for biological production, i.e. the growth of photosynthetic plants and performance of dependent animals in the food chain. Regional climate sets the base-line, but within this there are local gradients of climate according to topography, especially as this determines altitude. As ground elevation rises, it becomes colder, wetter, windier and cloudier, and this produces the sequence of zones of vegetation showing decreasing stature, as described in the Introduction. Mean temperature falls by 1 °C for every 150 m increase (or by 1 °F for every 270 ft) so that the tops of mountains rising, say, 915 m (3000 ft) above low plains are some 6 °C (11 °F) colder on average than at their foot. During 1913–1940, the Ben Nevis instruments at 1342 m (4400 ft) recorded an annual average of 248 days with minimum temperature below freezing point, compared with the 62 days in Fort William, virtually at sea level, at its base.

Rainfall also tends to increase with altitude, but the heaviest rainfall sometimes occurs to the leeward of the highest peaks. In Lakeland a rain gauge at Sprinkling Tarn in the Scafell Range regularly gives the highest rainfall recorded for the district, with an average of 4445 mm (175 in). Cloud cover also increases with altitude, and higher hill ranges are often notoriously misty places. Sunlight naturally decreases in parallel, and average windspeed rises. Indeed, the strong winds of the high ground produce an additional chill factor which enhances the prevailing coldness, particularly for warm-blooded creatures. The high mountain climate is distinguished not only by the considerable amount and frequency of snow-fall but also by the length of time during which the accumulated snow lies into the spring and summer, especially where it has drifted deeply in sheltered places from snow gathering grounds on extensive plateaux above.

Curiously, perhaps, these deeper snowbeds give protection to underlying vegetation from extremes of frost, and even animals may benefit. Ptarmigan and Red Grouse can live under the snow, digging tunnels, from which they can reach the underlying vegetation for food. When the snow has finally melted away by the following summer, there are often the accumulated patches of chopped-up grass collected by field voles for food and bedding along with their droppings, and in Scandinavia the snowbeds are a winter refuge for lemmings. By contrast, on exposed ground of the high tops, the snow is either blown clear or left as a thin layer which soon melts out. The unprotected ground here is subject to fluctuations between intense frost and thaw, and this alternation of conditions over a long period produces soil and stone movements which form the variety of patterned ground so characteristic of the high Arctic. On our higher summits there are the equivalents of stone polygons and networks or soil hummocks on level ground, which become elongated into stripes and ridges as the ground begins to slope.

Some high slopes are thrown into staircase systems of terraces, of widely varying size, resulting from gradual slumping during time of thaw of the ground which has become puffed-up with ice crystals. Related features are 'wind-rows' of vegetation running in parallel bands around the contour. Sometimes severity of the wind removes finer particles of soil, which build up new deposits in more sheltered spots and leave an armour of surface stones appearing as though steam-rollered in. Stones may also be tilted on their sides by movement, and secondary re-distribution of fine material by wind and water often occurs. The evidence of instability is prevalent but, while some of the smaller-scale features are still active, the really large ones belong to ancient times when ice sheets were all around and conditions had a truly Arctic severity ('periglacial'). The vegetation of these montane summit areas is mainly moss or lichen heath, carpeting the ground in some places but in others reduced to a scanty covering on stony and gravelly fell-fields: the summer haunt of Dotterel and, lately, Purple Sandpiper.

Steep-sided mountains show aspect effects, since slopes facing between north-west and east see rather little of the sun in our northern latitudes, and so tend to be damper and chillier than those which are sun-exposed. In some mountain areas this influence appears to have resulted in more intense ice-action on the shady aspects, resulting in a prevalence of corries and deep hollows facing between north and east. The layers of loose rock debris which are exposed on many high summits, and lie as unstable scree on the mountain flanks, have been formed by frost-shattering and erosion over long periods, but especially during periods of colder climate. In a few of the high late snow-hollows, they are the nesting place of the Snow Bunting, with a tiny fringe population in this southern outpost from its main Arctic range. It is here a living Ice Age relic.

This leads us to consider the other crucial aspect of the mountain climate: the decrease in length of growing season with altitude, and the general effect of the marked seasonal differences in availability of food for most birds. There is a repetition, though on a smaller and less dramatic scale, of the pattern of interplay between environment and bird life in the Arctic regions. A short summer produces a passing phase of bounty, sufficient to draw birds which can exploit it for breeding, but followed by decline to a food shortage requiring the early retreat of all but the hardiest to more favourable wintering grounds elsewhere.

Most of the hill birds, or at least the passerines and waders which figure conspicuously, are invertebrate feeders. These are the type which most noticeably depart from the uplands before winter really takes hold and reappear again when it gives way to the following spring. The invertebrates become largely unavailable, taking refuge from the cold in deep hiding places or over-wintering as earlier phases of their life-cycle. The Wren is one of the few invertebrate feeders which can survive the hill winter, at least at lower elevations, by foraging in the crannies of block litters and walls, and among long vegetation such as old, leggy heather. The Dipper continues to work the hill streams, and appears to have problems only during the most intense and prolonged frosts. The Stonechat, by contrast, clings to the moorlands at its peril and, even in the lowlands beyond, often suffers terrible losses during prolonged, hard winters.

Most of the other smaller passerines simply leave, the Meadow Pipits, Skylarks and Twites for the lowlands, and the Wheatears, Whinchats and Ring Ouzels for warmer shores beyond Britain and Ireland. The waders, which probe beneath the soil surface for food as well as taking some from above, virtually all quit the high ground at the end of summer, or even before. Their time on the moors is short, often no more than four months at most.

The vegetarians are few, but are the species best fitted for a year-round residence on the hills. Their most notable members are the Red Grouse, so typical of heather moorlands all over Britain, and the Ptarmigan, which replaces it at higher levels in the Highlands. The Grouse feeds mainly on ling heather, but also eats a variety of other dwarf shrubs growing plentifully on some moors. The Ptarmigan also crops the more alpine dwarf shrubs and their berries. Black Grouse feed on a variety of plants, but have a special fondness for the buds and foliage of trees and shrubs. Insects are important in the diet of Grouse chicks, so that even the vegetarians are not wholly independent of animal food. Many invertebrate feeders also take some vegetable food. One of the few birds to seek the hills in winter is the Snow Bunting, arrived here from the Arctic. It ranges the uplands, usually in little parties, in search of the seeds of rushes and grasses, and appears undeterred by snow cover, so long as the heads of these plants remain visible.

No hill bird is adapted to a diet of the grasses which are often dominant in the vegetation, so that the grasslands tend to be less productive for birds than the dwarf shrub heaths. Grey geese of one species or another here and there resort to lower boglands where, with their strong necks and bills, they pull up and eat the succulent bases of white beak sedge and cotton grasses. The White-fronts and Pink-feet are winter visitors to Britain, and are among the few species attracted by this inhospitable terrain at that season. The Greylag nests very locally and sparingly on moors in the Hebrides and northern Highlands, though feral populations have become established through introductions in southern Scotland and parts of England.

The few other birds which are able to subsist on the mountains and uplands for the whole year mostly belong to the small group of scavengers and predators. The Raven and the Crows, Carrion and Hooded, can find enough carrion, especially of sheep or Red Deer, and other pickings, to last them through the winter. In some areas, at least, Peregrines are within reach of a winter food supply sufficient for them to cling to the nesting haunts except perhaps in times of heavy snow. The Golden Eagle and Buzzard usually benefit from both food sources, taking live wild prey or carrion according to availability. The large size of some of these birds is an advantage, in that periods of enforced fast are less taxing for them than for smaller predators. The Raven and Golden Eagle seldom desert the hills, however hard the going. By contrast, the Merlin usually leaves the moors in late summer along with the Meadow Pipits, Skylarks and Wheatears that are its food supply. The Kestrel also tends to seek the moorland edge and the marginal land, where feeding is better during winter, and many young of the year disperse to distant lowland haunts.

The other birds which make up the upland breeding avifauna, comprising mostly a miscellany of waterfowl, are also species largely excluded from these habitats in winter by cold and shortage of appropriate food. Autumn sees a modest influx of new species to the uplands (Massey, 1978). Passage migrants such as Green Sandpiper move through in small numbers. Of the winter visitors there are flocks of Fieldfares and Redwings which work the fellside rowans, hawthorns and bilberries for their fruit, but then move into the lower country as winter advances. Woodcock also pass through the hills during autumn, and the numbers of Snipe and Short-eared Owls wintering on the lower uplands are boosted by incomers from abroad. A few Rough-legged Buzzards establish winter haunts among our hills, small numbers of Goldeneye take to the lower upland lakes, and Jack Snipe spread out sparingly in marshy places.

The weather markedly affects the habits of the birds which stay on the uplands during autumn and winter. If it is mild, there is little change from late summer. In some hill districts, moderate numbers of Snipe and Mallard remain, small parties of Golden Plover may be seen, and predators such as Merlin, Hen Harrier and Short-eared Owl hunt the lower moors. With heavy snow and prolonged frost, all these birds disappear, and most of the others come lower. The resident raptors and scavengers tend to hunt more over the moorland edge and the upper farms, and some may move out of their high crag roosting haunts temporarily. Some hill birds seek shelter in the hill woods. The Red Grouse often congregate on the moorland edge and even the Ptarmigan may descend into submontane ground.

Severe weather also delays the return of the migrants and partial migrants to the uplands at the end of winter, sometimes by a month or more. The earliest to reappear – Golden Plover, Lapwing and Curlew – often come and go from the uplands according to the vagaries of the weather during February and March. Heavy snow or hard frost will send them down to the upper farms, or even back to the low country beyond, there to re-form into flocks. As Chapman (1907) so aptly described, the return of spring to the northern moorlands tends to be a much delayed and erratic process compared with its advent in the southern lowlands. This has been especially the case since 1960, with frequent heavy snowfalls in late March and April, and icy winds blowing continuously from the east for weeks on end. Then, eventually, the rain, wind and cold subside, and the moorlands suddenly become alive with bird song and activity in the warm sunshine, as the feathered occupants make up for lost time in following their nesting impulses. Many show a burst of territorial activity, and the various waders go through their captivating display flights, while the Grouse strut and crow. These magical spring days on the moors are often short-lived, as our oceanic climate reasserts itself, and we – as well as the birds – have to make the most of them.

The bird breeding season in the uplands is spread out between the beginning of March and mid-June, as regards the onset of laying for the earliest and latest species (Table 1). Many species do not return until April or even May, by which time some of the earlier nesters already have young. Each has a breeding cycle that has evolved to fit the particular circumstances under which the species lives. Not surprisingly, the high montane and far northern birds tend to be the latest nesters, for they belong to the places where the return of spring is the most delayed.

Climate has these direct and present effects, but its influence is also indirect and historical, and to appreciate fully how it has moulded both the habitat and the associated bird fauna, we have later to look at its past action.

The importance of geology, topography and soils

The form and elevation of the hills have a strong influence on their bird life. The strictly montane species are limited to mountains of sufficient altitude, and these are mainly in the Scottish Highlands. Yet the highest hills are not necessarily the best: Ben Nevis is a relatively poor mountain for birds. The rugged and precipitous ranges are naturally much better habitat for the cliff-nesters than gentle moorlands with few rocks, where availability of suitable outcrops may severely limit the numbers of such birds. Yet

Fig. 2. The voice of spring on the moorlands: Curlew in display flight.

sharp peaks and narrow watersheds are often rather unproductive, and gentle contouring usually gives the greatest diversity and numbers of mountain birds.

The Boreal and Arctic birds which in Britain make up the true upland avifauna are mostly species that in their northern continental haunts breed especially on flat or gently sloping tundra and fell-field, or in marshes and beside lakes and rivers. Most of them nest down virtually to sea level in the far north, and their British occurrence mainly as upland species reflects the restriction of equivalent habitats to higher elevations in these southern latitudes. The waders and waterfowl are especially well represented, though the passerines form quite a strong element. Apart from the Common Sandpiper, which sometimes nests on fairly steep banks, nearly all the British waders need ground with a slope angle of less than 15 ° for nesting. The reasons are unclear, for though there is an obvious connection with their dependence on wet or moist habitats for feeding, this need applies also to species which feed a good deal on drier ground, such as the Lapwing and Golden Plover. The chicks, which leave the nest soon after hatching, may be less well adapted to feeding on steep ground. A consequence is that, for example, the broad, rolling moorlands of the Carboniferous formations in the Pennines and Cheviots are ideal terrain for the waders, whereas they are poorly represented on the mostly steep fells of Lakeland.

The montane species, such as Ptarmigan, Dotterel and Snow-Bunting, are found particularly on the more massive high mountains, where high plateaux give extensive feeding and nesting terrain for the first two, and their adjoining hollows favour retention of late snowbeds, which seem to attract the last. Yet some of the characteristic hill birds occur over a wide range of topography and/or altitude. Areas looking somewhat similar in their structure and relief can also sometimes differ a good deal in their bird populations; and some species are missing from many apparently suitable-looking areas. Other factors controlling distribution are evidently at work.

The chance occurrence of ancient geological processes and irregular distribution of hard, resistant rocks have produced the uneven scatter of mountain systems in Britain and Ireland. Varying properties of the different rock formations and differential glacial action also account for a wide variety of mountain forms across the two countries. No less significant is the influence of geology on soil formation and, in particular, the ability of the soil to support living organisms. This 'carrying capacity' is a measure of the intrinsic fertility of the land, and hence the amount and nutritive value of the vegetation cover on which dependent animals feed. In the hill country, the over-riding influence which determines this productivity is the amount of available calcium in the soil. Vertebrate animals need a minimum supply of calcium for the development of their bony skeletons, but the more important effect of the mineral is its alleviation of soil acidity, measured as pH. It has become the custom in Britain to regard soils below pH 4.5 as distinctly acidic, and those between pH 6.0 and 7.0 as fairly basic.

Calcium is usually the base nutrient most freely taken up by the soil exchange system, so producing high pH. It is derived especially from lime (calcium carbonate or calcite) as the most freely available form in the parent rock. Calcareous rocks, such as limestone, some mica–schists and calcite-bearing sandstones, shales and igneous rocks, and also marine sand with abundant molluscan shell fragments, produce calcium-rich soils in the range pH 6.0–7.0. Some, containing solid particles of lime, are alkaline, with pH over 7.0. Other rocks, such as basalts, dolerite, hornblende and olivine schists and gneisses, and andesite, often contain a good deal of calcium, but in a less readily available form, and give rise to less strongly calcareous soils. At the other end of the scale are hard, siliceous rocks containing very little available lime, such as quartzite, Millstone Grit, Torridonian sandstone, granite, rhyolite and granophyre. These yield soils with low base content and low pH.

The soils of wet climates, such as that of western Britain and Ireland, are much affected by the washing out of the soluble nutrients which counteract acidity. This process of *leaching* partly redistributes these nutrients to lower depths in the soil but tends to remove large amounts in stream water, or produces enrichment of lower ground by the balancing process of deposition by *flushing* (Pearsall, 1950). The rocks from which the soils were formed are themselves mostly hard and acidic, yielding a small amount of nutrients, and the low temperatures of the mountain climate further slow the processes of chemical weathering which release these substances from the parent rock. The mountain environment thus almost conspires to produce soils which have a marked tendency to high acidity and, hence, infertility. The acidity has the effect not only of reducing the supply of essential nutrients to living organisms, but also of inhibiting the microbial activity which controls the breakdown of organic material and the mobilisation of other essential nutrients so important to animals, such as nitrogen and phosphorus. The more acidic soils, below pH 4.5–5.0, thus tend to be biologically inactive (the absence of earthworms is a simple indication) and show a tendency towards surface accumulation of undecomposed organic matter, as raw humus.

Another feature of our cool, oceanic climate with its heavy rainfall and high atmospheric humidity is the tendency to waterlogging of soils. This feature reduces decay of dead plant remains and leads on flat or gently sloping ground to extensive development of the deep peat of raised and blanket bogs, which at an early stage insulates the living surface from the ground below and makes it dependent on atmospheric fall-out for essential nutrients. Such peat is also thus characterised by low pH and low fertility.

Acidic, nutrient-poor soils usually support vegetation which grows slowly, giving small biomass production, with a low content of carbohydrate, protein and mineral nutrients in its tissues. It accordingly supports a low biomass of dependent animals, and usually a limited species variety. Wild herbivores have an innate ability to recognise the more fertile soils with nutritious vegetation, and will seek out such areas and graze them preferentially, if they are localised. Britain and Ireland suffer, in fact, from the unhappy chance that their mountainous regions not only sustain a highly unfavourable climate for soil formation, but are also composed largely of hard, non-calcareous rocks. Soil fertility in the hill country is generally low, and more productive ground is at a premium. This has led to wide differences in stocking rates for various kinds of managed grazing animals, notably sheep, according to these differences in productivity.

All this has enormous significance for the bird populations. Vegetarian birds respond to differences in soil fertility in much the same way as herbivorous mammals. They selectively eat the vegetation of better soils, and here achieve higher population density and breeding performance than on the poorer soils. This has been convincingly demonstrated for the Red Grouse (see p. 96). Ptarmigan also achieve higher densities on the rather richer schist mountains around Drumochter and the Devil's Elbow, than on the sterile granite of the high Cairngorms; and the effect also extends to the invertebrate feeders. The Golden Plover is a characteristic bird of the acidic heather moorlands and cotton-grass blanket bogs, yet the densest breeding population now known is on a grass-covered limestone plateau in the Pennines. In many areas, the adults nest on the moorland but feed especially on the improved grasslands of the upper farms. The Dunlin, mostly sparsely distributed over wet flows and high, peaty watersheds, reaches quite spectacular breeding densities on some of the calcareous machair marshes of the Hebrides. Dave Parmelee tells me that, in the North American Arctic, occurrences of limestone have marked effects in producing locally high breeding densities of various wader species, which otherwise occur sparsely over non-calcareous areas of tundra and fell-field.

The influence of the past: the Ice Age and subsequent changes

The foregoing sections deal with some general ecological principles which are basic to an understanding of upland birds. The more detailed aspects of this bird fauna – its composition, distribution, abundance and status – can only be revealed by an historical examination of the environmental changes which have occurred from the end of the last Ice Age up to the present day.

Ice Ages are both climatic and geological processes, illustrating the interaction of both in the moulding of the landscape. The distribution and form of mountain systems in Britain are largely a matter of the geological chance that old, hard rocks occur mainly in the north and west, where glacial processes have also been most intense. In Ireland, such rocks are distributed mainly around the edge of the land mass. The last glaciation was, however, merely the latest in a long and continuous series of episodes in geological time, which have created the mountains we see today. Long periods of accumulation and compression of ancient marine sediments were followed by massive folding and fracturing of the rocks, and then erosion and reduction of the upthrust highlands by water and atmospheric processes. Interruptions by volcanic activity both above and below the ground surface gave an irregular scatter of hard, resistant rocks. In the Highlands, especially, enormous pressures and heat also transformed sedimentary rocks into metamorphic types, some of similar hardness to the igneous rocks.

The last Ice Age was thus a phase of further sculpturing and polishing amongst mountain systems whose features were already well formed. The work of previous glaciations became enhanced: the corries deepened, the high peaks and ridges sharpened, the U-shaped valleys and truncated spurs further accentuated, and the waterfall ravines below the hanging valleys deepened. The previous mantle of soils and rock debris was largely scraped away and redistributed by the ice, leaving the land with a fresh covering of the materials from which soils are formed. Ice and frost greatly enhance the grinding and shattering of rocks into ever smaller particles, and their subsequent transport and deposition in morainic and melt-water deposits. The chemical weathering of rock material, which is a crucial part of soil formation, increases with temperature and so becomes more important in the warmer periods following glaciations.

Virtually the whole of Britain north of the line from Glamorgan to Teesmouth, and all but the southernmost parts of Ireland, were covered by ice at the maximum advance of the last (Devensian) glaciation. From the record obtained by the analysis of pollen grains and larger plant remains preserved in deposits of glacial material, lake and river sediments, and peat, historical ecologists have reconstructed the nature of the vegetation as it was at the time of maximum glaciation and then the sequence of its subsequent change as the ice sheets receded northwards and upwards. The record is supplemented, though in a far more fragmentary way, by the remains of animal bones, including those of some birds, which chance to have been preserved and found, mainly in cave earths. The story is a fascinating one, and admirably described by Godwin (1975).

At the height of glaciation, from about 50 000 up to 13 000 years ago, the more northerly and mountainous parts of Britain evidently had something of the appearance of Greenland as it is today. Much of the land was buried under ice, but higher mountain systems broke through and in summer, at least, showed areas of ground free of snow, though with feeder glaciers in their deeper corries and valleys. In sheltered and low-lying places towards the sea in the west, there may well have been ice-free ground with vegetation and some birds and other animals. But eastwards there was no coastline, for so much water was locked up as ice in the northern hemisphere that sea levels were 100 m lower than today, and Britain and the rest of Europe were a single continuous land mass. It seems that the Irish Sea was always deep enough to prevent Britain and Ireland from being joined, but there may have been narrow land connections. To the south of the ice limits there were large tracts of arctic–alpine tundra and heath, then steppe or open scrub, while true forest lay farther beyond, across the land bridge which much later became the southern part of the North Sea.

The tundra zone had a mixture of wetter habitats in the valleys and poorly drained plains, and drier types on the higher or more sloping ground. Shallow lakes and marshes evidently abounded, and open water, swamp and fen vegetation predominated here, with sedges, reed and other grasses, herbs and small willows. There was probably much wet tundra with a patchwork of low scrub formed especially of dwarf birch, low northern willows and other small shrubs, and abundant mosses. Dry ground had a low and often open vegetation with arctic–alpine grasses, sedges, other small herbs such as saxifrages, dwarf azalea, least willow, and mosses and lichens. On such drier terrain frost-sorting maintained raw, open soils in various kinds of surface patterning, such as the networks and stripes still visible in fossil form in the Breckland today. In places there was a good deal of juniper and, perhaps, scattered small birches, but no real woodland.

In upland districts of southern England and the Midlands, and southern Ireland, the landscape probably came to resemble that of some of the high Norwegian table-lands today, such as the Hardangervidda: a mixture of rivers, lakes and marshes, with extensive low scrub tundra, rising to stony fell-fields on the higher and drier ground. In hillier places there would be long-lasting snowbeds in summer, while the higher mountain systems of the north and west would be largely regions of permanent ice and snow.

There is little evidence, in the form of actual remains, of the birds which lived in these arctic–alpine habitats south of the permanent ice. Harrison (1987) lists numerous bird species whose bones have been found in caves in south-west Britain and dated to the last glaciation or its end phase. Although these include at least 12 Arctic or Boreal breeders, he points out that the evidence of association with cold climate is equivocal, since most are migrants and the remains are often mixed with those of temperate zone species. Bones from Chudleigh Cave on the east side of Dartmoor include those of the sedentary Ptarmigan and Willow/Red Grouse, as well as Snow Bunting, Shorelark and Lapland Bunting; so that the Dartmoor region may well have had a good Arctic-type avifauna at some period in the last glaciation. Beyond this, we can only draw inference by looking at comparable situations and vegetation types widespread in the Arctic today. It seems a fair assumption that there was some similarity to the present bird community of the treeless tundras of Lapland.

The waders (or shore birds) are perhaps the most characteristic group of tundra birds, thriving during the summer on the myriad insects and other lowly animal forms of the extensive wet areas created by spring thawing above a deeper layer of permanently frozen ground. There would most probably have been species of the fell-fields and mountain-marshes: Dotterel, Purple Sandpiper, Golden Plover, Dunlin, Little Stint, Temminck's Stint, Jack Snipe, Turnstone, Red-necked Phalarope, Broad-billed Sandpiper, Bar-tailed Godwit and Whimbrel. These are birds well represented in the European Arctic today, and most of them occur far south in the Scandinavian mountains. Others which belong to Siberia or the North American Arctic such as Grey Plover, Sanderling, Grey Phalarope, Curlew Sandpiper and Knot, could also have occurred, but less certainly. Ptarmigan and Willow Grouse would represent the galliforms, while a variety of wildfowl doubtless frequented the rivers, lakes and marshes, including Whooper Swan, geese – which could have included all the species wintering in Britain today – and an assortment of northern ducks: Long-tailed, Scaup, Common Scoter, Velvet Scoter, Wigeon, Pintail, Teal, Goosander, Red-breasted Merganser and, possibly, Smew and King Eider. The Divers probably included Great Northern as well as Black-throated and Red-throated, and Slavonian Grebes would have been another lake-nesting species. The passerines would have included Snow Bunting, Bluethroat and perhaps Arctic Redpoll and Red-throated Pipit in the scrubbier and damper places. The predators could well have numbered the Snowy Owl, Gyr Falcon, Rough-legged Buzzard, Merlin, Raven, Arctic Skua, Long-tailed Skua and possibly Pomarine Skua.

The Arctic-type bird fauna may have contained more species than this, or it may have had fewer. Some which belong now to this habitat in northern Europe may never have been drawn to nest in Ice Age Britain or Ireland. Much would have depended on their total distribution and migration patterns at that time. Most of the nesting birds must have been only spring and summer visitors, as they are in the Arctic today. Winter return of snow and ice forced them southwards again, even though the lower latitude would have shortened the period of seasonal cold, under the same regime of day length we know today. Willow Grouse and Ptarmigan probably remained on the same ground as they do in the far north. The bird fauna of the tundra zone in those far-off times must have been an exciting one. And it had the company of some spectacular mammals, in the Mammoth, Woolly Rhinoceros, Bison, Giant Deer, Musk Ox, Lion, Brown Bear, Horse, Reindeer, Red Deer, Elk, Spotted Hyaena and Wolf.

Around 15 000 years ago, the climate slowly became warmer, causing the southern edges of the ice and the lower snowfields to melt and retreat northwards and upwards. Great spreads of bare ground were formed from material left by the glaciers as boulder clay and morainic land forms, or as outwash plains and other melt water deposits. The raw, open soils and shallow lakes were colonised by plants from the adjoining tundra zone to the south, so that this gradually advanced northwards. The birds and other animals would have followed suit, so that the whole Arctic ecosystem migrated slowly northwards, and was followed sequentially by the zones of steppe, scrub and forest which lay to the south and east. The scrub was mainly of juniper, willows, hazel and birch, and the first woodland was dominated by birch and hazel. Scots pine followed later, and there was evidently a belt of mixed forest similar to that of the taiga forming the northern fringe of Boreal coniferous forest lying across Eurasia. The Scots pine is the only large conifer which spread into Britain and Ireland after retreat of the ice, so that our Boreal forest is less varied than in continental Europe. This would have brought new birds which belong to the Boreal forest region: Fieldfare, Redwing and Brambling, for sure, but quite possibly Waxwing and Arctic Warbler, with waders such as Wood and Green Sandpipers, Spotted Redshank, Great Snipe and Greenshank in the extensive swamplands that persisted within the spreading forests.

The ice continued to retreat and dwindle within the mountainous districts, and the wave-like spread of main vegetation zones followed in its

Fig. 3. Snowman of the mountain tundra: Snowy Owl on breeding grounds.

Fig. 4. Changes in climate, habitat and bird fauna since the end of the Ice Age.

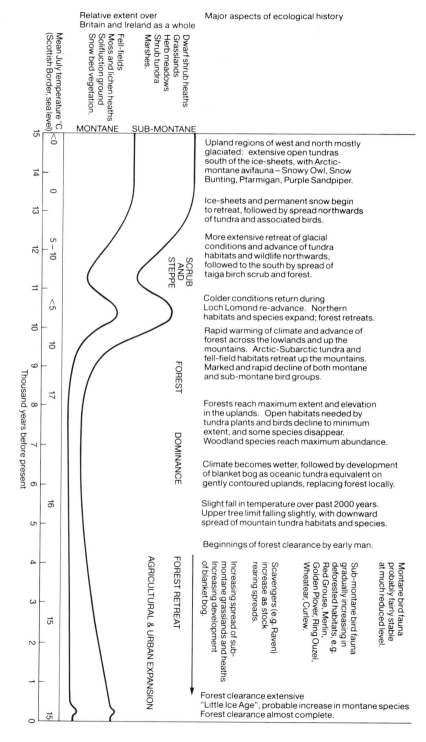

wake. The Boreal period saw a predominance of pine forest, but with increasing spread also of mixed broad-leaved woodland, of oak, ash, wych elm, aspen, small-leaved lime, hazel, holly and rowan. Forest came to occupy nearly all the lower drier ground, though swamps developed in many valleys and on waterlogged plains. The trees spread gradually up the mountain slopes, displacing the open tundras and fell-fields to ever higher levels. Forest appeared to reach its maximum altitudinal advance in the uplands, representing the period of greatest warmth, between 6000 and 5000 BC (Birks, 1988). Soon after, the warm, dry conditions changed to the markedly wetter climate marking the onset of the Atlantic period. Alder spread rapidly, and the wetness of climate allowed widespread initiation of acidic peat development in badly drained situations. The bogmosses (*Sphagnum*) flourished and there was a general beginning to the growth of the great expanses of blanket bog which so characterises our oceanic western and northern regions.

In many places, as recent erosion of the peat has revealed, bog development replaced the forests. The numerous buried stumps and roots of trees, especially pine and birch, show that forest cover was general on many moorlands now conspicuously treeless, and reached higher levels than would be possible under the present climate. In the Cairngorms, the remains of sizeable pines occur in eroding peat bogs up to at least 800 m, whereas surviving fragments of stunted pinewood representing the present climatic tree-line on exposed western slopes are at only 640 m. The indications from various lines of evidence are that the upper forest edge in most hill regions stood several hundred feet higher than its climatic limit today. The northwards extension of ivy, mistletoe, holly, hazel and the Pond Tortoise beyond their present limits suggest that mean summer temperature was probably 2 °C higher in northern Europe in the mid postglacial, compared with today (Godwin, 1975).

This upward extension of forest had important ecological consequences. Many lower hills and mountains must have become completely forested, except where the ground was too wet, or precipitous and rocky. A scrub zone above the forest would have further restricted mountain tundra and fell-field to the highest plateaux and summits, mainly in the Scottish Highlands. The ascent of woody vegetation evidently caused the extinction of many montane plants, both in particular districts, and from the British mountains as a whole. Open tundra birds would suffer similar depletion, and some high latitude species probably disappeared altogether from our mountains, while others became restricted to the limited occurrences of suitable habitat. Ptarmigan, Golden Plover and Red Grouse were probably far less numerous and widespread in the hills than at present, because their open habitats were so much less extensive. The subsequent cooling which became marked (by 1 °C) around 2500 years ago forced the upper edges of the forest downwards again, and allowed the open montane habitats to re-expand. Conditions for the non-forest birds thus became more widespread again. Since then, average temperatures in Europe have shown a further long-term downward trend (by another 1 °C) up to the present, but climatic change proceeds in a step-wise fashion, with fluctuations which vary on a time scale from annual to centennial (Lamb, 1982). Temperatures are now believed to be rising again (p. 224). Upland birds respond in different ways to this varying time pattern of change in climate.

Climate has always varied geographically at any one time, and the upland bird fauna has been much influenced by the spatial pattern of climate which long ago became established across Britain and Ireland. Even within the space of 11° of latitude and 12° of longitude, there are major gradients of climate. The south to north decrease in mean temperature interacts with the east to west increase in oceanicity. One of the most marked effects has been a substantial depression of the altitudinal zones of hill vegetation in a north-westerly direction, shown especially by the tree-line and the lower limits of montane dwarf shrub heath. Another is the great increase in the development of deep peat bogs in the same direction, first the raised bogs of the lowland plains and then the upland blanket bogs which reach their greatest European extent in western Ireland and northern Scotland. Some upland birds noticeably follow this downward shift in 'life-zones' (e.g. Golden Plover, Dotterel, Ptarmigan, Merlin), emphasising again that altitude limits are of rather little use in defining either the uplands or their birds.

Human impact on land use

While deteriorating climate was depressing the tree-line, human activity was increasingly removing forest cover and creating open moorland at all levels. Forest clearance began earlier in the south and may have been significant by the end of the Neolithic period, some 3000 years ago. By the Middle Ages little woodland remained on the hills of England, Wales and Southern Scotland, and by AD 1800 even the Scottish Highlands were largely deforested. The main aim was to create grazing range, but the pattern of upland land use has diverged through a variety of reasons. One important factor has been the differences in intrinsic carrying capacity depending on the nature of the parent rocks and the fertility of their derived soils (see p. 24). The intensity of use, and the impact on vegetation and fauna, has depended especially on the ability of the land to support either domesticated or wild animals which have been managed as a resource.

The human settlers in the hills have tended to exploit the better ground preferentially, through their animals, and have sometimes ignored the poorest areas. They have also, over the centuries, tried in places to improve soil fertility by adding essential nutrients. The crofting lands of western Scotland and Ireland were won from the rocky heaths and peat bogs by laborious carrying of shell sand and seaweed from the shore. The 'reclamation' of deforested lower slopes and moorland edges to form the hill farms has everywhere involved a substantial enhancement of soil fertility, by the addition of lime and also manure, for nutrients such as nitrogen, phosphorus and potassium are especially important to animal production. Manipulation of grazing pressure, involving also treading and dunging effects, has been much used to give improvements. Modern work on hill land reclamation and improvement necessarily involves enhancement of nutrient levels by fertiliser application, but may also include ploughing, herbicide treatment and reseeding with commercial grass strains.

Pearsall (1950) has given a valuable summary account of the history of pastoral development in the British mountains and moorlands. Much of the deforested hill ground found early use as sheep-walk, and the expansion of hill sheep-rearing and growth of the wool trade from the twelfth century appears to have been closely associated with the Cistercian monasteries in northern England, parts of Wales and the Tweed valley. In other regions, especially Wales and Scotland, cattle rearing was predominant, with goats important in places, and large herds of Red Deer ranging

Fig. 5. Abernethy Forest and the Cairngorms, Inverness-shire. Remnants of the Boreal forest, of native Scots Pine, rising to the high montane zone. Forest nesting habitat of Golden Eagle and, in larger clearings, Greenshank.

wild over higher and remoter ground. The Highlands, in particular, remained an undeveloped region, with more woodland surviving, and smaller scale grazing of cattle, goats and sheep. The extraordinary episode of the Highland Clearances changed this, from the end of the eighteenth century until around 1830. Whole crofting populations were evicted from their native glens to make way for great flocks of sheep, managed by a few shepherds and 'vermin' hunters. Many of the remaining forests were removed and grazing land expanded further. Birds and mammals regarded as predators of sheep and lambs were relentlessly killed.

In northern England, the end of the eighteenth century already saw an increasing interest in managing some moorlands for Red Grouse to be shot in due season for sport. By around 1840, it also appeared to many Highland landowners that more reward was to be had by exchanging interest in sheep for the sporting return from management of the native Red Deer or Red Grouse. Thus was set the Victorian fashion for the wealthy to buy Highland and other hill estates, build shooting lodges, and employ deer stalkers and grouse keepers. The slaughter of predators intensified and was conducted with a ruthless efficiency, as the horrific records of various estates testify. Virtually the whole of the Highlands was parcelled out into such game reserves, together with large areas of the Southern Uplands, Northern England and the Welsh hills. By 1900, nearly all the uplands of

Fig. 6. Vale of Maentwrog, Snowdonia. The native woods of oak and birch in the valleys of North Wales give some idea of the more general appearance of the upland landscape before the advent of Man. Before these woods were extensively cleared, there was much less scope for the upland birds which need open habitats.

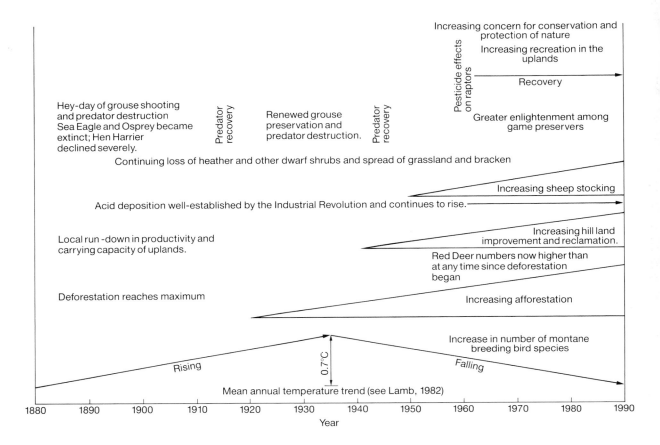

Increasing concern for conservation and protection of nature

Increasing recreation in the uplands

Pesticide effects on raptors

Recovery

Hey-day of grouse shooting and predator destruction Sea Eagle and Osprey became extinct; Hen Harrier declined severely.

Predator recovery

Renewed grouse preservation and predator destruction.

Predator recovery

Greater enlightenment among game preservers

Continuing loss of heather and other dwarf shrubs and spread of grassland and bracken

Increasing sheep stocking

Acid deposition well-established by the Industrial Revolution and continues to rise.

Local run-down in productivity and carrying capacity of uplands.

Increasing hill land improvement and reclamation.

Red Deer numbers now higher than at any time since deforestation began

Deforestation reaches maximum

Increasing afforestation

Increase in number of montane breeding bird species

Rising

0.7°C

Falling

Mean annual temperature trend (see Lamb, 1982)

1880   1890   1900   1910   1920   1930   1940   1950   1960   1970   1980   1990

Year

Fig. 7. Ecological changes in the uplands over the period 1880–1990.

Fig. 8. Altitudinal zonation and extent of major habitats before and after human occupation of the uplands. (*a*) The uplands before Man: a general cover of forest up to 600–650 m, except for lakes (1), bogs (2) and cliffs (3). At the upper edge of the forest was a zone of subalpine scrub, passing into higher-level heaths, grasslands, marshes and an uppermost zone with late snow-beds, solifluction features and high montane vegetation.

(*b*) The uplands at present: most of the forest has gone and the lower slopes have become hill farms. The former forest bog has become smaller and the edges are now artificially defined by draining. Fragments of native forest have survived but artificial shelter woods have been planted. The deforested ground is covered with varying proportions of dwarf shrub heath, grassland and bracken, according to the prevailing management regime.

Britain were run as sheep-walk, deer 'forest' or grouse moor. Forest was
reduced to scattered fragments, and the almost universal grazing and
burning largely prevented regeneration of the trees.

In most of the Irish mountains, forest clearance became even more complete by AD 1800. Fine hill oakwoods survive in Kerry and Cork, but elsewhere there has been a general replacement by a range of dwarf shrub
heath and grassland similar to that in Britain. Extensive blanket bog formation in many parts of the west, especially Co Mayo and Galway, may have

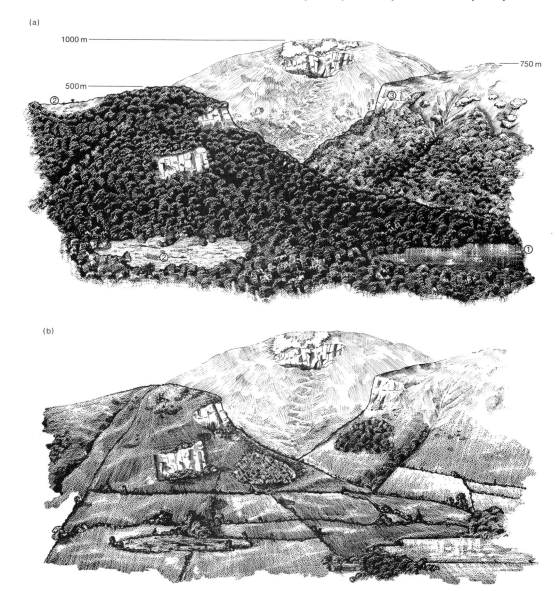

been a more natural termination of forest cover. Grazing range for cattle
and sheep was the main form of management, and though there was local
sporting interest in Red Grouse and Red Deer, the Irish hills have become
predominantly sheep-walks.

These man-made developments have had quite profound effects on the
bird-life of the hill country. They represent the most recent phase of post-
glacial change in environment, following and further modifying the effects
of the natural processes of climatic shift and geological activity. We shall be
much concerned to understand the importance of these and other still
more recent human impacts on the birds of the hill country. Chapters 2–7
give an account of the upland bird fauna according to six major types of
upland distinguished by important features of topography, geography and
land use. These subdivisions are arbitrary and, to some extent, overlapping, but give a useful classification of uplands for the purposes of this
book. Table 1 gives miscellaneous information for the 66 upland species,
while Table 2 summarises breeding distribution in Britain. Other sources
of information about upland ecology and wildlife include Pearsall (1950),
Ratcliffe (1977) and Usher and Thompson (1988).

2

The sheep-walks

While sheep are run on hills virtually throughout upland Britain, the districts regarded especially as the hill sheep-walks are: Dartmoor and Exmoor, most of the Welsh mountains, the Pennines, Lakeland Fells, Cheviots and Southern Uplands. The southern and western Highlands have much sheep ground, but interest in Red Deer is often important in this region, while grouse moors still represent a major concern farther east. The Irish mountains are mostly managed for sheep. The sheep-walks are typified by the smooth and rounded, though often steep-sided, hills of the sedimentary formations of central Wales and most of the Southern Uplands. Formed here of relatively soft shales and greywackes (gritstone) of Ordovician and Silurian age, they are rather little broken by large crags, and exposed rock is more often minor outcrops with patches of scree (Fig. 9). Typically, their massifs are penetrated by deep, branching and water-worn glens, often with rocky stream ravines and waterfalls. Here and there is more rugged ground, where glaciation has carved rocky hollows or steepened the valley sides into high escarpments.

In other districts, especially where igneous or metamorphic rocks occur extensively, much more rugged hills prevail. In Snowdonia, Lakeland, Galloway and much of the western Highlands and Islands, and western Ireland, the mountainous sheep country has a strikingly bold relief, with numerous high cliffs, sharp peaks, narrow ridges and deep-cut corries, all features associated with pronounced glaciation in hard rock areas. These districts are popular with climbers and hill walkers and contain some of our most beautiful upland scenery (Fig. 10).

In some areas the earlier forests were replaced directly by grassland, but in others, dwarf shrubs (notably heather and bilberry) became dominant. When the dwarf shrubs are heavily grazed over a long period they are gradually replaced by grasses, or often by bracken on dry ground. The process is accelerated by repeated moor burning. The sheep-walks have thus come typically to be dominated by grasslands, of sheep's fescue, bents, wavy hair grass and vernal grass on drier soils, and mat grass, flying bent and heath rush on moister types. In many grassland-clad uplands, heather

39

is almost confined to ungrazed cliff ledges, or the crowns of occasional huge detached blocks which cannot be reached by sheep. The widespread retreat of heather moorland even over the past 40 years is confirmed by comparison of older and recent air photographs. Bracken beds are often extensive on dry hillsides and have also extended considerably.

Many sheep-walks still have areas of heather, but often in poor condition and clearly retreating. In some districts, heather and bilberry ground has persisted most extensively on areas of particularly acidic rock with infertile soils, or on especially rocky and precipitous terrain. Such areas have discouraged the high sheep stocking rates which have built up on the better grazing land. Wetter ground with peat is also usually of lower quality for sheep than dry terrain. The most productive areas, on calcareous and other basic rocks, are everywhere green and grassy, with a larger variety of grasses and small herbs forming nutritious swards.

Woodland composed of native trees is patchy and often sparse, but many districts still have remnants on the lower hill slopes. Usually these are of oak or birch, but on the more fertile soils there are ash, wych elm and hazel as well. In some areas, almost the only surviving examples are fringes in deep-cut glens and along the side of rocky stream courses, or fragments on broken scarps (Fig. 16). Alder fringes on stream alluvium and wet soils are characteristic, and some hillsides have numerous old

Fig. 9. Blencathra, Lake District. Sheep-walks of acidic grassland, heather, bilberry and bracken, on Skiddaw Slate. Enclosed pastures and hay meadows of the upper farms below, which with the lower slopes of this 870 m hill would once have been wooded up to around 500 m.

bushes of hawthorn. Rowan is widespread, and scattered growths on cliffs at 450–600 m are often the nearest approach to a tree-line. Holly and willows are plentiful in many areas, and small trees or tall shrubs of more local occurrence are bird cherry, juniper and blackthorn. The common gorse forms thickets on some lower hillsides, and in parts of Wales, south-west England and Ireland the smaller, western gorse is abundant either in grassland or mixed with heather.

In most sheep hill areas there are frequent small plantations made for shelter, as belts or clumps in exposed situations. They are mostly of conifers, especially Norway spruce, Scots pine and larch, but in some areas sycamore was planted a good deal, especially around the farmsteads, and some fine old trees often occur. Regeneration of woodland on the open sheep-walks is mostly prevented by grazing, so that many remaining woods are effectively moribund, unless appropriate measures are taken to restore them. Seedlings, especially birch and rowan, usually come up quickly if sheep are removed for a while, but this happens rather seldom, and woods are often valued as winter shelter and feeding ground for the animals.

Birds of prey and the crow family

Let us start with birds of the hills in central Wales, where the old counties of Cardigan, Carmarthen, Brecon, Radnor and Montgomery come

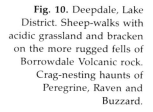

Fig. 10. Deepdale, Lake District. Sheep-walks with acidic grassland and bracken on the more rugged fells of Borrowdale Volcanic rock. Crag-nesting haunts of Peregrine, Raven and Buzzard.

together. It was in the sequestered valleys of these uplands, with their hanging oakwoods, where the RED KITE found its last refuge in this country. A once common scavenger in our medieval towns, the Kite lost ground when urban hygiene improved, and its familiarity with Man became its undoing as the great age of game preservation took over (Fisher, 1949). A bird especially vulnerable to the keeper's wiles, it suffered a headlong decline, disappearing from most of Britain by 1850. It hung on in the remote headwaters of the Rivers Tywi, Teifi and Wye, and after a precarious existence for many years, slowly spread out again to recolonise lost ground over a wider area of Wales. The region where it persisted has a distinctive ecological character, but there are other areas in Wales, south-west and northern England and the Southern Uplands sufficiently similar for one to look for other explanations of survival. The relevant specialness of this central Wales stronghold seems to be the fortunate circumstance that

Fig. 11. The last refuge: Red Kite in its upland oakwood and sheep-walk terrain, where it nests in trees.

there was here one of the very few large enough areas of suitable habitat with a negligible interest in game preservation.

Persecution by other countryfolk must have contributed to decline, for the Kite would, by feeding on dead carcases, be suspected of killing sheep and lambs. It also had the reputation of being a taker of domestic fowl. The security of the bird in its last refuge in central Wales was thus only relative, and by 1900 the Kite was clearly heading for early extinction in Britain. The very rarity of the bird was proving an irresistible magnet to egg collectors. In 1903, largely on the initiative of Professor J.H. Salter, but with both local and national assistance from concerned ornithologists, a campaign fund was established to protect the remaining Kites. The subsequent story is of a long, difficult and often unrewarding struggle against heavy odds. It has become an undoubted success for bird conservation though, from the slowness and restriction of recovery, not an unqualified achievement.

By 1903 only 4 nesting pairs were known, but in 1914 J. Walpole-Bond estimated that there were 10–12 pairs. Up to 1950, there was seldom if ever complete census information, which led to a tendency to under-estimate numbers. During the period 1900–50, the population evidently hovered between at least 5 and 10 pairs, with perhaps a small number of non-breeders in the area. Davies and Davis (1973) and Davis and Newton (1981) have documented the Kite population from 1950 onwards, and most of my account is drawn from these valuable studies. During 1951–60, annual averages of 10 pairs bred, rearing 8 young, with 7.7 non-breeding birds. During 1961–72, the annual average had risen to 23 breeding pairs, 15 young reared, and 11.7 non-breeders. Increase has continued slowly, and in 1989, 52 pairs bred, 32 successfully, rearing 47 young, and about 85 non-breeding birds were present (P.E. Davis, unpublished).

During this expansion, the nesting range has steadily radiated outwards, but a great deal of apparently suitable country remains unoccupied. Kites occasionally appear in hill country in England and even Scotland, yet the only proved nesting outside Wales has been of single instances, in Devon (1913) and Cornwall (1920). Why is there failure to reoccupy more ground, now that there is a modest breeding population? Perhaps other apparently similar areas do not satisfy the precise habitat requirements of the Kite. Or possibly reproductive output is simply too low to produce a big enough surplus of birds, and most of these potential recruits go to filling gaps in the established breeding population. Sufficient surplus young Kites may leave the main breeding area, but then become mopped up by a heavy mortality as they move into areas where the chances of survival are less. Another suggestion is that the population became so small that it is now suffering from genetic defects associated with inbreeding.

The Kite's claim to be considered as an upland bird rests largely on its dependence in the Welsh haunts on the open hill ground and ffridd, as well as the marginal land and enclosed fields, outside the nesting woods. This is the preferred feeding habitat, and the bird ranges widely but variably in its hunting, showing only a loose territorialism compared with the Buzzard. Pairs defend their nest sites, but otherwise seem to tolerate each other's proximity, and may overlap a good deal in their feeding movements. In winter, they sometimes flock to some especially favoured feeding place, such as a refuse tip where 20+ birds have been seen at once. Kites are associated with a somewhat old-fashioned type of hill farming, which results in plenty of sheep and lamb carrion, and encourages large numbers of corvids, including Carrion Crows, Jackdaws and Magpies, whose newly fledged young then provide another food source for the Kites.

In an analysis of regurgitated pellets (Davis and Davis, 1981), 69% contained sheep remains, and nearly 58% had remains of other mammals, mainly Mole, Rat, Hedgehog, Short-tailed Vole, Rabbit and Brown Hare, plus a mixture of at least 18 other species. Bird remains varied from 44% frequency in spring and summer to 25% in autumn and winter, covering at least 21 species, and including especially corvids and Black-headed Gulls (obtained from nesting colonies at moorland tarns). Observations of food remains on nests and direct predation increased the number of mammal species to 24 and birds to 44 (including Wood Pigeon, Lapwing, domestic fowl and sundry passerines) and added frog and toad. A variety of invertebrates are commonly taken, though their bulk contribution to total food appears to be small.

This food spectrum suggests that, unless there are areas where improved sheep husbandry has greatly reduced the supply of carrion, a large total extent of hill country elsewhere should be suitable feeding habitat for Kites. Many parts of the Pennines, where bird populations are relatively large, should be especially favourable. The recent establishment of a few breeding areas in more lowland farm environments in Wales also implies that a large total area of country is potentially available to the Kite in Britain. Most upland areas have at least patches of woodland, or shelter belts which could provide nest sites. Habitat availability thus does not appear to be limiting.

Reproductive output is relatively low. With a mean clutch size of 2.2 eggs, mean fledged brood size of 1.4 young and mean productivity of 0.5 young per nesting pair (Davis and Newton, 1981), there is not exactly the capacity for a population explosion. Disturbance by humans, including egg collecting, still contributes to low breeding performance. There are still risks to nesting adults, especially from poisoning. Yet those who have studied Kites closely for many years believe that survival of the established adult breeders in their Welsh stronghold is now reasonably good, and find no evidence suggesting adverse effects of inbreeding. The indications are that some of the young fledged disperse from the main breeding area each year, but others remain, eventually to settle as breeding adults. A few Kites evidently arrive here from the continent, but they probably represent only a small temporary addition to the non-breeding population.

While an increased population pressure from a larger surplus of young birds would no doubt enhance the chances of successful colonisation elsewhere, the clear implication is that the main resistance to the spread of the Kite is the hostile reception which these wanderers meet sooner or later. Recoveries of ringed Welsh birds have been made in several parts of the English lowlands, and the Winter Atlas shows 8 recent winter sightings in Mid and East Anglia. The fate of those Kites which sojourn in the lowlands can only be surmised. Several have been reported as 'found dead' in the game-rearing areas of southern England. For the RSPB, Cadbury, Elliot and Harbard (1988) report that during 1971–87, 3 Kites were shot, 17 poisoned and another 11 suspected of being poisoned: the numbers were equally distributed between Wales and England. One which made its way to Galloway a few years ago was shot by a farmer, and one of the 31 RSPB records was for a Kite killed in Northern Ireland. Latest news is that, between 11 March and mid June 1989, 10 Kites were found poisoned, all but 2 by fenthion, in their main breeding district of south-central Wales (P.E. Davis, unpublished).

The slow recovery of our Kite population seems likely to continue under the present regime of protection, and its happier present position is a tribute to the devotion of all those who have worked so hard for its survival. We must applaud this as one of the few notable successes of the twentieth century conservation movement, as an actual gain and not just a holding of the status quo. Yet for it to become a resounding success in the longer term may require that this country learns to treat its wildlife more kindly. Two commendable attempts to re-introduce Red Kites to Scotland and England in 1989 have been met by the daunting news that some of the birds in both release areas have rapidly dispersed to far distant places (P.E. Davis, personal communication).

The Kite shares its nesting haunts and feeding range with two other large bird predators and scavengers, the BUZZARD and RAVEN. Both are especially numerous in central Wales, and give added interest to this district. All three of these fine birds may often be seen soaring at once over the rolling sheep-walks of the Cambrian Mountains. The Buzzard and Raven are nowadays more truly hill birds than the Kite, in that they flourish in quite treeless country and nest up to higher altitudes, at 650 m or more, on crag ledges. These were also once widespread tree-nesters in lowland England, and the Buzzard retains a strong outpost in the New Forest. In Wales and the West Country, besides the rock-nesters, large tree-nesting populations of both species still occur where there are woods and shelter belts, from the lower hills and valleys down into the marginal land and even lowland farms. In places the Raven and Buzzard have a more or less continuous distribution, from the hills across the lowlands to the rocky coasts, where both are equally at home on the sea cliffs.

The explanation for this uneven distribution is somewhat similar to that for the Kite. From an earlier almost continuous occurrence over Britain and Ireland, there has been a retreat through the relentless persecution associated with game preserving, reinforced by the hostility of sheep farmers. The Buzzard and Raven have disappeared from much of the lowlands, but held the line much better than the Kite, especially in the west. Their cliff-nesting habits and ability to live farther from human presence in remote uplands were probably advantageous when persecution was severe. Moore (1957) published almost exactly complementary maps of the distribution and density of gamekeepers and of Buzzards in Britain; and this matching presence and absence was interpreted as cause and effect. As attitudes to raptors have improved, there has been a welcome tendency for Buzzards to regain lost ground in some parts of Britain, spreading back into typical lowland habitats. Yet they still have low chances of survival in areas where game birds are preserved, and in the hill country they are often conspicuously absent as breeders on or near grouse moors. In the sample survey of breeding distribution for 1983 reported by Taylor, Hudson and Horne (1988), Buzzards were found to have increased slightly within their main range in parts of western Britain and more substantially along some of the edges of their range, compared with 1968–72. There was, however, little sign of any general eastwards expansion. The bird is now at a very low ebb in the Pennines and Cheviots, and in Ireland it has made little recovery, being found mainly in a limited area of the north-east.

Buzzards from the west certainly make their way into the more eastern hills and lowlands in the autumn and winter, but there run high risks of a fatal contact with unfriendly people. Picozzi and Weir (1976) recorded a total of 223 Buzzards killed on just four estates in north-east Scotland in 1968, so it is small wonder that no increase is reported from that region. By contrast, in the Langholm district of the Southern Uplands, keepering was relaxed in the 1970s, and the first pair settled soon after. Numbers rose rapidly and by 1985 had reached 29 pairs. After this a particularly zealous keeper and his assistants set to work again, and in their combined area there was a decline from 16 pairs in 1985 to 3 pairs and a single bird in 1989. At the same time, they also effectively prevented the establishment of Golden Eagle and Goshawk in the area. Over the other part of the district, all 13 Buzzard territories of 1985 remained occupied in 1989 (Ian Newton, personal communication).

Other problems for the Buzzard have concerned food supply. The rabbit is its favourite prey, though young rather than adults are mostly taken. The Buzzard formerly reached it highest breeding densities in areas where rabbits swarmed, such as parts of mid-Wales. Then, in 1953–54, came myxomatosis, and within two years this important food source had virtually vanished over much of the country. Moore (1957) reported the impact of the myxomatosis epidemic on the Buzzard population. A good deal of non-breeding occurred at first, and in some districts numbers thinned out markedly. Yet

Fig. 12. Breeding distribution of Buzzards in Britain and Ireland during 1968–72 (from Sharrock, 1976). The three grades of evidence of breeding – possible, probable and confirmed – are shown on the maps as three increasing sizes of dots.

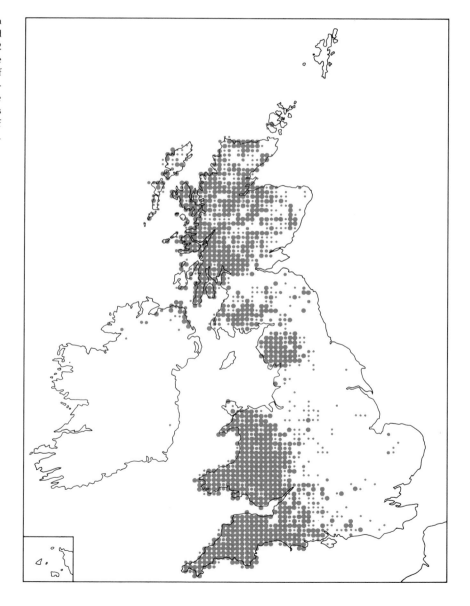

it soon appeared that Buzzards in some areas were adjusting their diet to cope with the absence of rabbits, by increased feeding on carrion, especially sheep, and taking of smaller mammals and birds (including nestlings and fledglings) was found. Holdsworth (1971) found that Buzzards around Sedbergh adapted to field voles as principal prey, and that their breeding success thereafter tended to follow fluctuations in vole numbers.

Numbers of Buzzards have held up remarkably well in many hill areas, and expanded into adjoining lowlands. They are perhaps a little lower than the maximum density before myxomatosis, having reached a new balance against the local food supply. Central Wales still has a high breeding density of Buzzards. Newton, Davis and Davis (1982) found an average of 24 pairs per 100 km^2 on upland sheep-walks, with an average distance of 1.13 km between an occupied nest and that of the nearest neighbour. Adjoining farmland had a still higher density, with 41 pairs per 100 km^2 and 0.87 km between nearest neighbours. Over a smaller study area on Dartmoor, P. Dare (unpublished) found 14–16 pairs in 33 km^2 and nearest neighbour spacing of 1.1 km. In the Spey Valley of Inverness-shire, Weir and Picozzi (1983) located 38 pairs in 173 km (22 pairs per 100 km^2) with nearest neighbour distance of 1.5 km.

Weir and Picozzi (1983) observed that the established breeding population was markedly territorial and sedentary, defending home ranges throughout the year by various flight displays and perching in conspicuous places, besides direct aggression. In areas of dense population, the owners of contiguous territories often display simultaneously, and up to five or more pairs may join to soar and circle in close company, then drift apart over their separate nesting grounds. On Speyside, defended areas were smaller in winter, but nesting territories, averaging 400 ha with some overlap, were fairly constant over four years. A few first-year birds established winter territories, but these were usually abandoned during the following spring, and only three new breeding territories were established in four years. Gaps appearing in the paired territory holders were usually filled by other adults, but surplus adults otherwise appeared to be mainly nomadic. The breeding population was evidently close to saturation level and limited by winter food supply. The Spey Valley Buzzard breeding area was surrounded by ground unoccupied and apparently unsuitable, through lack of nest sites, scarcity of food or difficulty of hunting, or continuity of persecution.

While the hill Buzzard population is generally stable, there have been losses in certain areas. In the years immediately preceding myxomatosis at least 14 pairs of Buzzards nested along the Eden Valley side of the Crossfell range between Hartside and Dufton, either in the fell-foot woods or in rocks higher up in the fells. The lower slopes then swarmed with rabbits, but this animal has almost gone. Numbers held up fairly well until at least 1960, though with smaller clutches, of 2 rather than 3 eggs. After that, pairs dropped out gradually until by 1985 probably only 3 territories were still occupied.

Buzzards have held their numbers over most of Lakeland, and their increase in partly wooded country within or around the district has more than compensated for any hill losses. One peripheral area has become much depleted, however. Before 1940, Greystoke Park, lying just eastwards of the Skiddaw fell-group, was probably the best Buzzard area in the district, with at least 12 nesting pairs in about 2550 ha. This was an area of limestone grassland which then swarmed with rabbits. During the Second World War, part of the area was acquired by the Forestry Commission. Most of the old nesting woods, in the form of scattered plantations, were felled, and the area was planted with some 800 ha of conifers. In 1986 only two pairs of Buzzards remained in Greystoke Park (G. Horne). The loss of nesting sites, feeding area and favourite prey have combined to bring about the decline.

In Galloway, the virtual disappearance of the once flourishing hill population of Buzzards is attributable largely to afforestation. The area between the old New Galloway – Newton Stewart railway, the Ken Valley, the Cree–Minnoch Valley, and the Ayrshire border, had at least 25 pairs of nesting Buzzards during 1946–55, some in hill bottom woods but most in crags, rather higher up. Rabbits were mainly around the edges of the hill massifs, and on these poor-quality sheep-walks, carrion mutton was a staple item in Buzzard diet. Afforestation 'took off' here after 1945, and the Galloway hills have been transformed from an area with a singularly low cover of woodland to a vast forest-clad landscape in which the bulk of the plantable ground is now blanketed with conifers (see Fig. 67, p.215). In 1987, Dick Roxburgh found one, or possibly two, pairs of Buzzards still nesting within this area. Afforestation has caused massive loss of food and feeding habitat. The ground above the forests provides little food, especially since the sheep have gone from much of it. New forest can sometimes provide plenty of nest sites where there were previously none, but the Buzzard's success still has to depend on the ability of the ground to supply food, and this may well depend on the extent of upland left unplanted.

Although they are so different in many ways, the fortunes of the Buzzard are closely connected with those of the RAVEN, its frequent neighbour in the hills. The Raven is above all others the bird of the sheep-walks. In Britain and Ireland, it is associated with sheep management virtually throughout its range. Some coastal Ravens scavenge a good deal from the sea-shore, and there is a thin scatter of pairs in some Highland deer forest country where sheep are few or absent. In most places, nevertheless, sheep and lamb carrion is the bird's staple diet (Ratcliffe, 1962). The disappearance of Ravens from southern and eastern England is not simply a matter of persecution. There was a close association in these lowland haunts with the former sheep-walks on open downs, extensive pasturage of a kind which has largely vanished in modern times. Ravens have also declined during the past 30 years in certain coastal areas such as south-west England, north Wales and Galloway. The explanation given by Mearns (1983) for Galloway probably applies throughout: that the food supply for these coastal Ravens has declined, through improved sheep husbandry (which reduces the number of dead animals) and steady replacement of the rough cliff-top grazings by improved pasture with cattle, or by arable.

The Raven is a fine bird, and to many naturalists, its presence greatly enhances the attractions of the hill country. Chapman (1924) viewed it as a survivor of a grander fauna, now mostly vanished into oblivion. I grew up on the edge of the Lake District, and under the influence of a group of energetic field ornithologists whose calendar was high-lighted each year by the approach of 'Raven time'. Plans for a round of nesting places were laid well in advance, and the weather watched anxiously as the end of February approached. In some years heavy snows would delay operations, and quite often forays into the fells would find the slopes frozen to an iron hardness, and the crags festooned with icicles and surmounted by snow cornices. Yet the Ravens were seldom much delayed in their nesting activities. The big stick nests, cleverly built into overhung and sheltered places in the rocks and snugly lined with sheep's wool, mostly had full sets of eggs by mid-March. Once the birds were sitting, further blizzards and frosts usually had no effect, and the naked young hatched some three weeks later into an often bleak and inclement world.

Fig. 13. Nature's scavenger: Raven on sheep carcase.

These nesting events are well-timed, nevertheless, as regards availability of food. Where sheep continue to range the higher fells in winter, a number usually come to an untimely end, from disease, malnutrition, exposure, burial in snow-drifts, or mishap on the crags. Their corpses, marked down by the patrolling Ravens, provide a source of food for some time ahead. Then around mid-April comes the annual lambing which, under the harsh conditions, is often attended by heavy losses, not only of lambs but also of ewes, whose strength has finally drained away. The placentas also lie around for scavengers to take. And so the young Ravens appear at just the time when food tends to be in fullest supply.

Not that it is necessarily easy going thereafter. The male Raven often has to work hard in providing for the whole family while the chicks are small and the female cannot leave them unattended. As it flies to and fro between carrion supplies and the nest, the bird's bulging pouch below the bill is clearly visible. The contents are regurgitated at the nest and the young fed when they signal hunger by begging for food. Ravens usually lay from 4 to 6 eggs, with 7 occasionally. The rate of infertility is usually rather low, so that the brood size at hatching is close to clutch size. Yet average *fledged* brood size is commonly well below average clutch size. There is a certain level of chick mortality, and this is related mainly to the availability of food.

The way in which this effect takes place was worked out by Lockie (1955*a*, *b*) for the Rook and Jackdaw, in which incubation also begins in earnest with the first or second egg. Since eggs are laid at daily intervals, this results in a staggered hatching of the clutch, typically over a period of several days. The young are thus of graded size and vigour, so that when food is brought, ability to beg for it varies similarly. The oldest and most vigorous chick is fed first and will continue to take precedence over its siblings until it is satiated and subsides. The next oldest then takes over, and so on until the whole brood is satisfied, for the time being. When food is scarce, the oldest chick is never satisfied and continues to compete with the rest of the brood. The youngest will usually be the first to 'feel the pinch'. Once a chick becomes so weak that it can no longer beg, it is not fed, and soon dies. So, depending on the amount of food brought to the nest, there may be a successive loss of the young chicks, until only one survives.

This mechanism for adjusting brood size appears to be characteristic of birds in which food supply at nesting time is variable, and the nestling periods prolonged. Natural selection favours survival of a smaller number of strong chicks rather than a larger number of weaklings. As a result, broods of one or two large young are quite common in the Raven, and more than four young are seldom reared. Fledged broods of five are quite rare. but I once saw a nest with six young, much of a size and somewhat over two weeks old, so that all had a good chance of surviving. Similar reduction in brood size also occurs in the Buzzard, which typically lays 3 eggs, but in some areas produces a majority of nests with only one youngster. There is some evidence (Newton *et al.*, 1982) to confirm the egg collectors' view that in hard winters which produce much carrion mutton, Ravens tend to lay bigger clutches than in mild winters when few sheep die.

The Raven is another strongly territorial bird in its nesting, and occupied eyries tend to be fairly regularly spaced out as a result (Ratcliffe, 1962). Breeding density varies widely and is probably highest now in parts of Wales. In the area of Central Wales with a high density of Buzzards, Newton *et al.* (1982) found a density of 21 pairs per 100 km^2 and nearest neighbour distance of 1.7 km. This was for upland pairs; in contrast to the Buzzard, Raven density was much lower in adjoining farmland. In this area tree and crag sites were about equally frequent. In North Wales, Dare (1986*a*) found breeding Ravens dispersed more or less regularly through Snowdonia at an average density of 10.5 pairs per 100 km^2, though in one area there were 12 pairs in 40 km^2. Mean nearest neighbour distance was 2.0 km. Nesting places here are more commonly in crags than trees. In Lakeland, overall density is only 7.5 pairs per 100 km^2, with nearest neighbour distance of 2.7 km (Ratcliffe, 1962; updated by G. Horne in 1987). Nesting here is almost entirely in crags.

Breeding density in parts of south-west England may perhaps approach that in central Wales. Elsewhere, Ravens are somewhat sparser in the hill country. Sometimes there is a lack of suitable nesting places, in both crags and trees, but some areas have plenty of scope for tree nesting, and here persecution is probably a limitation. In other districts it appears more likely that shortage of food is the main problem, and that variations in breeding density relate closely to those in carrying capacity. In Snowdonia, the population counted by Peter Dare during 1978–85 had increased by 80% since I surveyed some of the same area in 1950–53. This increase evidently reflects a substantial increase in sheep stocking rates. In parts of Lakeland, numbers of nesting Ravens are also higher than during the period 1900–60, though the increase is less than in Snowdonia. Sheep numbers have risen here, too (Table 3), but there has also been a recent trend towards wintering the flocks and holding them for lambing on the low ground, combined with increased winter feeding and diet supplements. These improvements in sheep husbandry have reduced the supply of sheep carrion, and the tendency for Raven numbers to drop back again locally may reflect this further change.

Decline and virtual disappearance of the Raven from the Pennines is almost certainly the result of Grouse keeping activity. The bird was never more than sparsely distributed here in recent times, but during the 1950s there were at least 10 more or less regular places between the Tyne

Table 3. *Increase in sheep numbers*

(*a*) Number of ewes in UK (millions)

1939	1971	1981	1983	1985	1987
12	14	15	16	17	21

(*b*) Number of ewes in Lake District National Park (thousands)

1945	1970	1975	1980	1985
200	232	265	287	322

Data provided by James Marsden, NCC.

Gap and Settle. Possibly no more than 1–3 pairs attempt to nest now and with very little success. In 1985, the first pair to nest for many years on the National Nature Reserve in Upper Teesdale was shot out before the eggs had hatched. Marked declines in the Cheviots and Southern Uplands are attributable to afforestation and removal of the sheep over large areas of upland. Once this food source disappears Ravens in affected territories usually stop nesting at once; they may hang around for a time but often drop out quite quickly. In this region, at least 100 pairs of Ravens nested regularly between 1945 and 1960. By 1974–76, only 48 of these territories were still occupied by pairs and 10 of these were not nesting (Marquiss, Newton & Ratcliffe, 1978). By 1981, the number of pairs had fallen to 35, of which only 23 were known to nest (Mearns, 1983).

Possibly 5 of these Raven pairs on small cliffs were displaced by rock-climbers, and another 4 by Golden Eagles, while a few more may have suffered food shortage through improved sheep husbandry. At least 31 pairs nevertheless dropped out because the new forests critically reduced their food supply. By contrast, in central Wales, afforestation appears to have had little effect on Raven numbers. Partly this may be because the plantations are smaller and more sheep ground remains in mosaic with the sheep-walks. There are also sheep living wild in some of the forests. The castings of Ravens living within or close to these forests still showed 90% frequency of sheep remains, though vole remains were more frequent than in pellets of birds farther from forest (Newton *et al.*, 1982).

Although Ravens in some areas have a rather high failure rate through egg collecting, and average fledged brood size is often only between 2 and 3 young, the total output of young birds from some of the main breeding strongholds is quite large annually. This is reflected in the size of many of the non-breeding flocks of Ravens which live in some upland areas and roost communally. Flocks of up to 60 birds are not uncommon, and in some coastal districts they may number up to 200 or even more. Established breeders appear to remain in solitary occupation of their territories throughout the year, though there may be temporary desertion during heavy winter snow and severe frost. They do not join the non-breeders, which often establish themselves in a locality with no occupied nesting place. Gaps in the breeding population are evidently filled by birds from these flocks, which sometimes appear to contain pairs, and evidently represent a reservoir of surplus birds. It is strange that they do not disperse and try to establish new nesting places elsewhere. Given the hostility to Ravens in many areas beyond the hills, they are, however, probably safer staying in parties among the uplands. Many young Ravens nevertheless disperse widely outside the breeding areas and suffer much increased mortality in seeking to colonise new country (Holyoak, 1971; Dare, 1986b). The Raven remains a widespread bird in the uplands, though it is scarce or absent in much of the grouse moor country. Its numbers can be expected to decline further as afforestation continues, especially in Scotland, and in its remaining strongholds, future status will depend greatly on what happens to upland sheep farming. Although we have come to regard the sheepwalks as a secure refuge, this is a bird whose fortunes have come to be closely tied to human use of the land, and thus to economic and political affairs.

Fig. 14. Female Peregrine feeding young. An eyrie in the Southern Uplands where Peregrine numbers have recovered as organochlorine pesticide contamination has declined. Edmund Fellowes

The mountain crags give refuge to another celebrated predator, the PEREGRINE FALCON. This bird has some claim to be regarded as the most exciting and glamorous of all our upland species. The Peregrine is a master of flight with powerful symbolic appeal. In the nesting season it lives especially in precipitous country of great beauty, and to many admirers its presence adds a fitting sense of completeness to some of our most dramatic scenery. The Peregrine breeds in the more rugged uplands throughout Britain and Ireland, but the sheep hills of Snowdonia, Lakeland, the Southern Uplands, the southern fringes of the Highlands, and certain of the Irish massifs support some of the highest breeding densities known anywhere in the World. It thins out markedly on the gentler moorlands, where there are few suitable cliffs, but readily resorts to man-made quarry faces.

There is now a large literature on the Peregrine, and the following brief sketch is drawn from major recent compilations (Hickey, 1969; Ratcliffe, 1980; Cade et al., 1988).

Though it occasionally takes ground animals, such as rabbits, the Peregrine is largely a bird feeder, killing its prey in spectacular headlong pursuit. Despite its catholic choice, with at least 132 different bird species taken in these islands, Peregrines tend to feed mainly on the dozen or so most plentiful kinds in the immediate vicinity. The Red Grouse is probably the favourite wild prey in areas where it is common, and a variety of upland waders is taken (especially Snipe, Lapwing, Curlew and Golden Plover) together with a selection of passerines ranging from corvids to Meadow Pipits. Waterfowl, especially ducks and gulls, are less important than might be supposed, but are taken by some pairs. In the sheep hills, where large wild prey is often sparse, the domestic pigeon is outstandingly the preferred prey item, freely available through the popularity of pigeon racing: it amounts to 70–85% by mass of total food in some districts.

In areas where availability of suitable nesting crags is not limiting, Peregrine breeding density is determined first by the strength of territorial interactions between adjacent pairs. There is circumstantial evidence that this spacing behaviour is itself more fundamentally dependent on the general availability of food in the neighbourhood. On the barren uplands of the northern and western Highlands, and western Ireland, where bird populations are sparse, Peregrine density is low and pairs widely scattered, averaging around 9.0 km between neighbours, and only 1–2 pairs/ 100 km^2. There are also historical indications that, in some of these districts, a long-term decline in moorland bird populations was followed by a thinning out of breeding Peregrines. By contrast, pairs remained more closely spaced in some nearby coastal locations where there were large numbers of seabirds. And in some of the more southern and eastern upland districts, where food supply (of both wild prey and domestic or feral pigeons) is good, Peregrines locally reach remarkably high breeding densities, with 4.5 km average distance between neighbouring pairs, and 4–7 pairs/100 km^2.

Whether Peregrines stay on their upland breeding areas during autumn and winter also depends on the food supply. In areas where winter food availability is good, established breeders are fairly sedentary, though they often appear to spend the day farther afield, and return to their home crags to roost. Choice of prey may also shift, with domestic pigeons less available, and Wood Pigeons, corvids, gulls and immigrant Fieldfares and Redwings more favoured. High mountain haunts in the Highlands are vacated for lower ground under the often arctic conditions, and some Highland Peregrines are said to feed mainly on Red Grouse and Ptarmigan during winter (D. Weir, personal communication).

Peregrines are not really high mountain birds. Most of the upland pairs nest in the submontane zone, between about 200 and 600 m. Curiously, the most elevated nesting places, at over 900 m, are on the coldest mountains, in the high, crag-lined corries of the Cairngorm area. Conditions here are usually still quite arctic in normal laying time, and the habitat looks more suited to Gyr Falcons. Further west and north in the Highlands, the upper limits to breeding fall quite markedly, and most nests lie between 100 and 400 m. The damp mistiness of higher elevations in the north-west appears to be more inhibitory than the greater cold of the eastern Highlands. Aspects appear to have little effect on selection of nesting places, given equal choice. The tendency for a majority of Peregrine cliffs to face between north and east in Snowdonia, Lakeland, the southern Highlands and many Irish mountains evidently reflects the greater glacial action, and hence frequency of high cliffs, on these aspects.

Peregrines tend to select the highest, steepest and most extensive cliffs available, with the greatest choice of nesting ledges safe from humans and ground predators such as Fox and Pine Marten. The biggest cliffs also usually have the most commanding outlook, from which to observe both prey and intruders and to gain advantage in attack. Some eyries are in situations that test the skills of the best rock-climbers. Many can be reached without too much difficulty by a lowered rope descent or by abseiling, but there is often an overhang which leaves the descender dangling free, requiring the additional awkward manoeuvre of working up enough pendulum motion to be able to grasp the rock and hang on. Some nests, especially in smaller crags, can be reached without ropes by even modest climbers, and in the less rugged areas there are a fair number in virtual 'walk-in' situations.

Peregrines have the same habit as Ravens and Buzzards in resorting to a selection of different eyrie ledges, often on separate cliffs, over a long period, though with the pattern of use varying widely between individuals and localities. All three species are closely associated in their occupation of the same crags and even – in different years – the same eyrie ledges. In those hill districts where Ravens are numerous, they nearly always share the same crags with Peregrines, and about one third of all falcon clutches are laid in unoccupied Raven eyries. Ravens will also build at times on disused Peregrine ledges. Sometimes the occupied eyries of the two are very close: less than 10 m apart on different sides of the same projecting buttress. There are limits to this tolerance, and Ravens will desert eggs if Peregrines decide to nest too close. Buzzards sometimes figure in site exchanges, and all three species have occupied the same ledge in successive years. Usually, though, Buzzards tend to occupy smaller and more broken rocks, leaving the bigger faces to the other two species. There is less tolerance between any of these three birds and Golden Eagles, and if Eagles should settle on their cliffs, the Peregrines, Ravens and Buzzards will usually quit and keep their distance subsequently. Yet Peregrines will occasionally lay in old Eagle eyries when the owners have moved to different crags.

The biggest cliffs in an area seem to exercise the same attractions for Ravens and Golden Eagles as for Peregrines. Their immediate visual appearance seems to trigger an instinctive common response in these crag nesters, that this is the best place to take up residence. Behind this response is the evolutionary fine-tuning of selection for those reactions which bring the greatest advantage to survival of an individual and its offspring in any situation. There is evident survival value in birds such as Peregrines, Ravens and Golden Eagles seeking out the biggest cliffs as their breeding stations, though the adaptation seems less marked in the Buzzard. Joe Hickey pointed out that the fidelity with which North American cliffs were held by Peregrines in the face of human interference – ranging from deliberate persecution to casual disturbance – was closely related to their height and remoteness. The smallest crags were the most readily deserted, and only held regularly where they were remote from human presence and access.

Exactly the same holds in Britain and Ireland, but the standards of acceptability for cliffs evidently vary between districts according to what is available. In less rugged districts, lacking big cliffs, quite small crags are often held regularly, even in the face of serious persecution. These ideas about suitability of cliffs are therefore relative, and there appears to be a trade-off in the Peregrine between using a place where there is a high risk of breeding failure (and also destruction of the breeders) or foregoing the chance to breed. Moreover, the annual use of the same, conspicuous cliffs may also increase the vulnerability of Peregrines and other predators to deliberate persecution by making the birds easy to locate. Choice of tall cliffs by no means gives security against modern Man armed with gun, trap and ropes, so that the survival value also is relative.

The persistence of the Peregrine in the face of long and relentless destruction is remarkable. There are many localities where gamekeepers are known to have killed one or both of a nesting pair annually over long periods, yet in each successive year another pair was in occupation. Sometimes replacement occurs within days by rapid re-mating of the survivor, and it has been known for the newcomer to become foster parent in incubating the eggs or rearing the young. At small and unsuitable rocks this kind of treatment is more likely to cause desertion, and once abandoned, such places may stay vacant for some years. Before 1940, many territories were also so relentlessly robbed by egg collectors that no young were reared in them over a long period, and yet the birds persisted in their fruitless efforts indefinitely. It seemed that in remote parts of the country enough young were reared annually to make good the losses in the nesting population elsewhere.

One of the other remarkable features of our Peregrine population up to around 1955, was its relative constancy. Before 1900, the records became scanty, but after this time, there was sufficient evidence from egg collectors, gamekeepers, falconers and others to indicate that in many parts of Britain and Ireland, the number of pairs attempting to nest varied rather little, either from one year or one decade, to the next. It had been noted, by writers back as far as the Middle Ages, that although there was often a good annual output of young, only the same number of nesting pairs was in occupation the next year. The majority of haunts were found to be regu-

Fig. 15. Graph of Peregrine decline and recovery in north-west England.

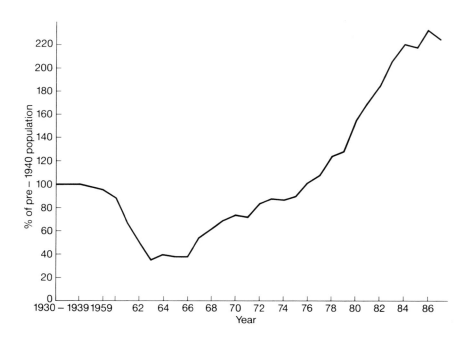

larly occupied every year, though the use of alternative cliffs in a territory can often require a good deal of searching to locate nesting pairs.

Its fidelity to conspicuous and traditional cliffs has also made the Peregrine one of the easier of our less common birds to census: when the BTO conducted the national survey in 1961–62, previous information on distribution was a valuable base-line. The results showed that there had been a dramatic crash in Peregrine population, beginning in southern England about 1956 and working rapidly in a wave-like spread northwards through Britain, and through much of Ireland. The cause was quickly diagnosed as the persistent organochlorine insecticides of agriculture, with secondary contamination of this and other predators along the food chain. Peregrines in many upland districts were severely affected even though local use of these chemicals was often slight. It was only in one district, the central Highlands, that a population remained relatively unaffected in numbers and breeding performance, evidently through the fortunate chance that this was an area both remote from pesticide use and supporting a sufficiently good year-round food supply to contain its Peregrines as a sedentary group.

By 1963, the national population was down to 44% of the pre-war level in Britain, and widespread breeding failure continued in this remaining fraction. A similar situation obtained in Ireland. During this year, however, restrictions on the use of the most toxic dieldrin group of insecticides as cereal seed-dressings appeared to take effect, and no further decline showed in 1964. There was little further change for a few years but, with additional restrictions on the use of the organochlorine insecticides, a slow recovery in Peregrines began to show. Monitoring of the population has traced its progress and spread, to the extremely satisfactory position whereby, around 1985, it could be said that the British Peregrine population was back to its pre-1940 level. The Irish population has also climbed back steadily to virtual normality.

This was a final vindication of the case that the persistent organochlorines were the cause of the decline, and the prediction that their reduction as environmental contaminants would be followed by a matching reversal in the trend. The dieldrin group were regarded as the main factor in the crash by producing a serious increase in adult mortality through their direct lethal effects. Although DDT was later shown to have substantial sublethal effects in reducing breeding performance, notably by causing eggshell-thinning, this appeared to be a secondary, though enhancing, factor in the decline. Dieldrin residues in Peregrines have declined to very low levels, and while those of DDT and DDE are still moderate in some areas, they have not been sufficient overall to prevent population recovery.

In many upland areas, the present position is the remarkable one of super-recovery, in which numbers have risen to higher levels than ever known before. In Cumbria, the hill breeding population during 1930–39 was estimated at not more than 34 pairs. In 1986 no less than 77 territories were occupied, and eggs were laid in all but 5 (G. Horne, P. Stott, T. Wells, T. Pickford and G. Fryer). In the Southern Uplands, the pre-1940 population was probably about 35 pairs. In 1987 there were at least 75 territory-holding pairs, of which at least 67 laid eggs (R. Roxburgh and G. Carse). In parts of the Pennines and adjoining fell groups, present numbers comfortably exceed pre-war levels, but I have been asked not to give details. There has been a surprising recovery in the Cheviots where the species was still barely represented in 1981. In 1989, the upland population in Wales was at least 173 pairs, compared with a pre-1940 level of only 65 pairs (RSPB). Numbers in Dunbartonshire and Stirlingshire increased from 17 pairs in 1981 to 22 pairs in 1986 (J. Mitchell).

The only upland areas where Peregrines have failed to recover completely or have even declined further are in the western and northern Highlands. In mountainous areas north of the Great Glen, numbers in 1981 were at 78% of the probable pre-1940 level. Poor food supply in this barren deer forest country, combined with a run of harsh, cold and wet springs, could have been responsible, but the reasons for this regional discrepancy remain uncertain. The coastal Peregrine population in the northern and western Highlands and Islands has remained depressed, with many vacant territories, single birds, non-breeding pairs and poor breeding performance. The cause is suspected to be contamination by marine pollutants (including the persistent PCBs) through feeding on seabirds. Despite the large output of young from the uplands and from coastal eyries elsewhere, Peregrines have not yet fully recovered in numbers along the south coast of England.

Elsewhere, the population levels which were previously regarded as a ceiling on Peregrine numbers have been far exceeded. There are two aspects to the increase. One is that there has been an increase in breeding density, with former territories of one pair becoming divided between two or even three pairs. The second is that increase has involved the occupation of small and unsatisfactory crags ('third-class') which were previously ignored. The first process has occurred especially in Lakeland and Snowdonia, and the second especially in the Southern Uplands and Pennines. Neither could have occurred without a plentiful supply of young Peregrines to fuel the increase, and this reflects not only a great reduction in pesticide hazards but also a decrease in direct persecution, as the result of the great efforts over Peregrine protection and public relations work. The good breeding performance in many areas is in contrast to the pre-1940 situation and recovery has steadily fed on itself. Increase in density would be less likely to occur if there had not also been an increase in food supply, and the growing popularity of pigeon racing may have had this effect in some districts.

The increased occupation of marginal habitats has included a considerable number of quarries, used as well as disused. These mostly have good rock faces, but levels of disturbance seemed often to be too high to draw Peregrines previously. Many of the new haunts are also tiny outcrops, or mere broken banks, often with 'walk-in' sites. My friend Dick Roxburgh has encouraged Peregrines to use several small rocks by making artificial scrapes looking exactly like the real thing. There have been several genuine ground nests on secluded moorlands, and sites on two old buildings in the hill country are recorded. Most interesting of all, at least two pairs of Peregrines in widely separated districts have used old tree nests of Ravens in which to lay.

The Peregrine story is thus one of the notable successes in bird conservation, so far. From the time, 25 years ago, when its rapid decline gave rise to gloomy predictions of possible extinction, it has recovered to reach unprecedented numbers in some districts. While it still has a hard time on some of the grouse moors, such as the Bowland Fells and North York Moors, in many upland areas the Peregrine has proved so adaptable in choice of nest site that there is considerable potential for further increase. If the new departure of tree-nesting became a firmly established habit, there would be still further scope for expansion into areas where the species has never been able to breed previously. In several upland districts it now outnumbers the Raven: a state of affairs once scarcely imaginable.

The KESTREL gives an interesting comparison with the Peregrine. A relatively common lowland bird, this is also the most widespread and numerous of the upland raptors. While subject to marked short-term fluctuations in numbers, its average population level far exceeds that of the Peregrine. From 10 to 20 pairs/100 km^2 is a fairly typical breeding density. Kestrels breed up to 670 m and hunt over higher ground, but they are largely birds of the submontane zone. In the uplands their staple food is usually the Short-tailed Field Vole, though other small rodents are taken as available, and small birds occasionally. A good many larger invertebrates are eaten, with beetles most in evidence, through the shiny pieces of wing-case mixed with vole fur in the castings.

Some hill Kestrels nest in trees, using unoccupied nests of Crows or other bird, but also in holes. Where available, cliffs are their preferred sites, and quite small outcrops or even broken banks are often used. Holes and crevices are used in preference to open ledges, though old Raven nests are sometimes occupied. Disused quarries are favourite sites and old buildings are sometimes used. Ground nesting occurs regularly only in Orkney, where mammal predators are absent. Perching places with droppings and castings occur on other rocks within the territory, at some distance from the actual nest. By the time incubation is advanced, their eggs usually lie on an insulating pad of the birds' own castings.

Kestrels usually keep their distance from occupied Peregrine eyries, for they are at some risk of becoming prey. Their feathers are by no means unusual at Peregrine plucking places. Presence of Kestrels at a crag is usually a sign that no Peregrines are breeding there. This is not invariably true, however, and Kestrels can quickly move into a crag where Peregrines have laid, failed and moved elsewhere. In one fascinating incident in the Southern Uplands in 1963, the Kestrels laid in the identical old Raven nest which had earlier held the falcons' eggs. The Peregrines then returned, displaced the Kestrels, took over and hatched their eggs, and reared all four chicks. I saw all six on the wing, the Peregrines angrily scolding against my intrusion, while the adopted young Kestrels now and then hovered in typical fashion. Three other instances of fostering of Kestrel clutches by Peregrines were known in the 1960s. This curious behaviour seemed to be associated with prior breaking of the Peregrines' own eggs, itself occasioned by organochlorine contamination.

During the 1960s, when Kestrels were showing severe decline in many lowland parts of Britain and Ireland as a result of organochlorine pesticide effects in agricultural districts, the upland populations were little affected. In contrast to the badly hit Peregrine, the diet and habits of the hill Kestrels largely protected them from contact with these toxic residues (Snow, 1968). Their large output of young probably helped to promote the general recovery in numbers in the lowlands when contamination levels were reduced.

Kestrel numbers are closely geared to those of Voles, and so tend to show cyclical fluctuations about a 5 year period (Snow, 1968). The changes are not always obvious, and seldom begin to approach those reported in various parts of the Scottish Borders during the great vole plagues of the late nineteenth century. Great increases in numbers of Short-eared Owls and Kestrels were then reported from the affected areas and breeding colonies of up to 18 pairs of Kestrels occurred in the Moffat Hills (Adair, 1892). Bolam (1912) later recorded breeding colonies of up to 20 pairs (with 7 nests in 40 m) of Kestrels along the crags of the Roman Wall in Northumberland which seemed to correspond with a local eruption in vole numbers. When the vole populations subsided, the Kestrels thinned out, returning to normal numbers and solitary nesting. Nowadays when vole numbers peak and their runs and tunnels appear abundantly in the hill grasslands, small groups of 3–4 nesting pairs of Kestrels can be found, with adjoining pairs only 200–400 m apart. The degree of colonial or clumped nesting during vole peaks evidently reflects the localisation of suitable nest sites: where sites are more evenly available, breeding dispersion remains more regular at the higher population densities.

Village (1982) found that in Eskdalemuir, Kestrels in summer defended an area immediately around the nest but shared the rest of their hunting range with other pairs. In winter, most pairs split up and adopted individual home ranges which were more exclusive, except at their edges. There was an overall population reduction in winter, and the general experience then is of an exodus of Kestrels from the hills. Many, and especially the young of the year, move southwards, while the others (probably including most breeders) become widely spread out over the lower moors, foothills and marginal land. The hills may be largely abandoned during severe weather, especially when deep snow hides the prey. Numbers increase again in the spring as the breeding population re-establishes its territories.

Two frequent associates of the Kestrel in the lower level and often smaller crags and old quarries are the Stock Dove and Jackdaw. STOCK DOVES occur in isolated pairs or small groups, nesting in rock crannies and holes on the face, or amongst large blocks. They also resort to disused buildings and hill barns. Sometimes they are in surprising proximity to Peregrine eyries, since they figure at times in prey remains. They are shy, little-known birds most usually seen beating a rapid retreat after being flushed with a clatter from some rock cranny. As vegetarians, they must find enough food in foraging about the lower hill ground, but they leave their upland haunts for more congenial places in the autumn, returning in February and March, and often occupying the same nest sites year after year.

JACKDAWS are usually in larger colonies, especially in cliffs with numerous crevices and fissures in which to nest. While they mostly distance themselves from Ravens and Peregrines, Jackdaws are sometimes found in close proximity to both these predators. Some groups breed where colonies of rabbits have riddled slopes with burrows, and others are in holes of older broadleaved trees on the lower hillsides. Most hill Jackdaws breed below 300 m and they, too, leave the more elevated haunts in winter. In Wales and northern England, Jackdaws and Rooks from the fellfoot colonies often join in mixed parties to range on fine summer days high over the hill slopes, apparently in search of insect food, such as craneflies, brackenclock beetles and antler moths. Jackdaws are regarded as nest predators by keepers and treated accordingly, but are mostly believed innocuous on the sheep-walks.

The CARRION CROW is a good example of a widely ranging bird well adapted to upland habitats. With its subspecific form, the HOODED CROW, which replaces the Carrion Crow in the Western Highlands and Islands, Ireland and the Isle of Man, this is one of the most widespread birds in the hill country. Absences usually point to persecution, and Crows are prevented from colonising many grouse moors, though there is no shortage of surplus birds from elsewhere to maintain the attempt. Should keepering become relaxed on any moors, the Crows will very soon establish and multiply.

The Carrion Crow is largely a tree nester and thus limited in its breeding mostly to levels below 450 m, though a few go as high as 550 m. Cliff-nesting appears to be more common in Wales than elsewhere, some sites being in places worthy of a Raven. Hooded Crows nest more often in rocks, and occasionally on the ground. Hill woodlands of all kinds are much favoured and the nests here tend to be less visible and in more difficult trees than on the open hill. Yet many pairs resort to the last vestiges of tree growth among the hills: the scattered thorn bushes typical of some lower slopes, the clumps and fringes of rowans, birches and alders persisting beside hill streams or on rocky places. Even isolated trees often hold the conspicuous bowl-shaped nest of sticks, lined like the Raven's with a thick pad of sheep's wool. These hill nests in small trees are very vulnerable, and many are robbed or shot out. But the Crow is a great trier, and even if one or both of a pair are killed, others usually take their places, probably from the

Fig. 16. Kielder Valley, Cheviots. Moorland stream habitat of Dipper and Common Sandpiper. Open fringe of birch and rowan with Carrion Crow nest, which can provide sites for Kestrel and Merlin. Rocky and heathery banks are typical habitat of Ring Ouzel and Wren.

non-breeding flocks so often to be seen. They will repeat at least three times a year, as successive clutches are lost, and often succeed in rearing young in the end. When the leaves on the trees are fully expanded, repeat nests can be difficult to see from any distance, and attempts can succeed in places all too obvious in their winter bareness. In parts of the Scottish Borders, nests are not uncommon in spindly little clumps of willow, little more than a metre above the ground.

Crows soon take to the few trees planted for shelter round hill shepherds' cottages, when these become abandoned. And in the few places where they are tolerated, Crows soon become bold, nesting in the gardens of occupied crofts in some parts of north-east Scotland. The nests are durable structures, and disintegrate only slowly over the years through attack by the elements. They are often rebuilt, sometimes in successive years. Old or failed Crow nests are often requisitioned by Kestrels and Merlins (Fig. 16), and a lone Scots Pine in a quiet Lakeland valley has several times, in different years, provided Crow nest sites for both species. Buzzards occasionally utilise them and may add sticks for enlargement.

The Carrion and Hooded Crows' diet overlaps a good deal with that of the Raven, but relationships between the two species are not well understood. Ravens will chase Crows from the vicinity of their eyries, and Crows seldom nest closer than about 400 m from an active Raven eyrie. Crows are usually all through the Raven country, however, and do not seem to be treated by Ravens as though they were obvious competitors. Crows take sheep and lamb carrion freely, but Ravens probably have first

Fig. 17. An unloved predator: Hooded Crow raiding Lapwing nest.

claim at any carcase. Crows are evidently much more systematic searchers than Ravens for the eggs and young of other birds. Some individuals take eggs to a stream or pool to eat, with the result that accumulations of the eggshells of grouse, wader and duck are sometimes found. Small mammals, birds, lizards and frogs are sometimes killed, and any dead animals could provide food. Yet Crows must also take a substantial amount of invertebrate and vegetable food, though perhaps variably according to season. They are omnivorous and can manage a living in hill areas which have the least variety in other birds.

Even in good Raven country, Carrion Crows are usually at far higher breeding density. Peter Davis finds them typically spaced at about 200 m intervals in the Kite country of Central Wales. In 1951, I found 16 occupied Crow nests in the Aber valley, Snowdonia, along a stretch 2 km long and average width of 0.6 km (1 pair/7.5 ha), with habitat varying from closed woodland to scattered trees. The Crows ranged outwards to an unknown distance, and the dense clustering of nests resulted partly from localisation of nesting sites. Interestingly, the clutches were all small, from 2 to 4 eggs. Crows are often patchily distributed because of the irregular occurrence of nesting trees, and it is strange that they have not adapted more fully to cliff nesting, which would increase the availability of sites besides giving greater security to nests. Hewson and Leitch (1982) found that in a 2025 ha coastal stretch of Argyll uplands, 20 pairs of Hooded Crows were concentrated along a tree-grown zone of 783 ha immediately above the shore, with a mean distance between occupied nests of 471 m and density of one nest to 39 ha. Non-breeding flocks are often seen during the nesting season, and evidently represent a population surplus to the established territory holders.

No bird is more universally detested by hill shepherds and gamekeepers. Crows are hardly able to kill adult sheep, but they have the notorious and gruesome habit of pecking out the eyes of any sick or injured animals, or those which are 'cast' and unable to stand again. Often the tongue is attacked as well. Ravens do this, too, but their attacks on sheep are less common. The reasons are still obscure, but one possibility is to secure the food supply: an animal which is blinded is as good as dead. Crows have the reputation of attacking even healthy lambs in this way, and there are few flockmasters who will tolerate their presence.

One result is the continuing widespread use of poison placed in dead animal remains or the eggs of chickens and ducks. Although this practice is now quite illegal, the end justifies the means to many keepers and shepherds, and the law is still widely flouted. Strychnine is less used now that more sophisticated modern poisons are available, notably phosdrin and alpha-chloralose. The indiscriminate use of these substances, purportedly against Crows and Foxes, kills many Ravens, Buzzards and Golden Eagles, and is one of the most serious causes of mortality in Welsh Kites. It even accounts for the occasional Peregrine.

Crows are undoubtedly serious nest and chick predators, and there is indirect evidence that they have contributed to the long-term decline of Red Grouse on some moorlands (pp. 98–100). The highest nesting densities of Golden Plover are mostly on grouse moors where Crows are strictly controlled. A worrying aspect of afforestation is its potential for bringing increased predation to moorland adjoining the new forests. From

Fig. 18. The charming Red-billed Crow: Choughs in a quarry haunt.

the thicket stage onwards, the plantations are colonised by Crows, whose nests are all but impossible to locate, and which forage widely over the remaining open moorland beyond. In areas such as Caithness and Sutherland, where the moorlands were almost treeless over large areas, and so afforded little nesting habitat for Crows, the new plantations are likely to increase exposure to these marauders for many of the remaining moorland birds. Crow problems probably become more serious through a scarcity or absence of the predators which could control them, such as Kite, Goshawk and Eagle Owl.

The CHOUGH is one of the most charming but also the most mysterious of our hill birds. It lives mainly on the sea cliffs, and nests inland only in north Wales, Islay, the Isle of Man and western Ireland. In Wales, almost all its inland breeding places are in man-made habitats: disused quarries and mine shafts. In the Isle of Man, Islay and Ireland, derelict buildings often afford nest sites. On quarry faces the nest is usually deep in a rock-cleft, and in old mines it is often in semi-darkness. Choughs may nest singly, or in loose groups, hardly colonies, with pairs spread out over a sizeable area. They are social birds, nevertheless, and join together to feed at all times of the year. Flocks are larger and most conspicuous outside the breeding season, and move around a district quite widely then.

The mystery is why Choughs are so highly localised and static in their nesting distribution. They were evidently once more widespread in Britain and Ireland, in both coastal and inland districts, and they have retreated over the past 150 years or so. Studies of their feeding habits point to a particular fondness for insect larvae and ants, and the need for short turf or recently burned heaths in which they can readily find their food, probing into the litter and soil with their delicate, curved beaks, so untypical of the crow family. It has been supposed that their decline or disappearance from some coastal areas is connected with the loss of these close-cropped and unimproved grasslands and heaths adjoining the cliffs, through agricultural improvements, and post-myxomatosis scrubbing over of many remaining areas (Bullock, Drewett and Mickleburgh, 1983).

This can hardly explain absence from Lakeland and the Southern Uplands, where there are still extensive grasslands and heaths apparently suitable as feeding habitat. Inland Chough breeding areas are all in areas of markedly oceanic climate, where winters are usually mild, with short-lived snow cover and rather little frost. And most of these inland birds are sufficiently close to the sea to retreat to coastal areas during prolonged periods of hard weather. Bullock et al. (1983) suggest that frozen ground prevents the birds digging for food. There is evidence for Chough decline during periods of unusually cold winters (Sharrock, 1976). This would, however, hardly explain the bird's failure ever to appear in some of these districts. The population on the Isle of Man looks across to the hills of Lakeland and south-west Scotland, but the bird is seldom seen here. It appears to be a species of conservative habits and movements are only local. The remaining inland populations are evidently fairly stable but delicately balanced. While there may be factors we do not understand, the Chough seems to exemplify a puzzling feature about many birds: they are essentially mobile creatures, potentially able to colonise new ground, yet the distribution of most species is markedly static and there is often an apparent resistance to spread beyond their well-established range.

Fig. 19. Breeding distribution of Choughs in Britain and Ireland during 1968–72 (from Sharrock, 1976).

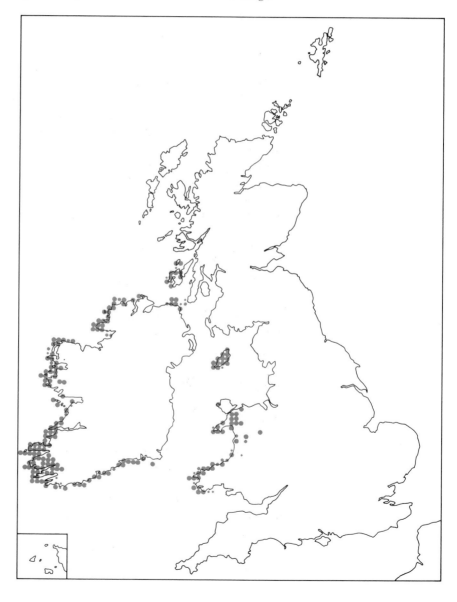

Song birds

The most ubiquitous upland bird is the MEADOW PIPIT. Like the bird's appearance, the song is unspectacular, and the Scottish name Moss Cheeper quite apt. The sight of little parties of Meadow Pipits drifting restlessly across the hills in late March is one of the earliest signs of spring here. The birds spread themselves out almost from sea level to near the highest summits at over 1200 m, occupying virtually the whole range of mountain and moorland habitats. Despite its abundance on the hill grasslands of Snowdonia, Seel and Walton (1979) found this a difficult species to study. The bird's general unobtrusiveness, especially the females, and the lack of song display by some males, made the counting of numbers tricky.

Pairs disperse into their separate territories, and Seel and Walton (1979) found that the home range of a breeding pair consisted of three concentric zones. The innermost zone represented the adults' normal living area, expanding into a middle zone as they satisfied their own requirements, and an outermost periphery corresponding to their food-collecting range in feeding the young. Within the Snowdonia study area, breeding density appeared to be relatively constant over a four year period, at a mean of 48 pairs/km^2. Other observers have found highest upland densities of 55–60 pairs/km^2, with little difference between the best of both grasslands and acidic moorlands, though some of the poorest grasslands and bog areas in the Highlands have no more than 5 pairs/km^2 (Sharrock, 1976). Density appears to thin out with altitude, too. Meadow Pipits breed in all hill areas of Britain and Ireland, and although the estimated total population of over 3 million pairs includes those in lowland and coastal habitats, this is still our most numerous upland bird by a wide margin. Even on the dullest grassland sheep-walks we can count on its faithful company, as one anxious pair hands us on to the next.

Not many hill Meadow Pipits have eggs before the end of April. In an analysis of a wide scatter of nest records, Coulson (1956) found that laying was earlier by about 5 days in the south than the north, and decreased with altitude by one day for every 40 m, evidently in response to increasing climatic severity. Average clutch size decreased from 4.52 eggs near sea level to 4.07 at over 305 m, and these differences existed in both first and second clutches. Coulson believed that at least half the pairs on his northern Pennine study area were double-brooded. Although the species lays appreciably larger clutches in Iceland and Norway, it is only single-brooded there, so the British populations have a higher reproductive potential. Nest failure decreased with altitude, which Coulson attributed to reduced predation at higher levels. Cuckoo parasitism accounted for 21% of clutch or brood failure in nests below 305 m, and especially below 75 m, but none of the record samples mentioned Cuckoo in nests above 305 m. This agrees with my own casual observations; in nearly 200 Meadow Pipit nests found at random, scattered over the British uplands and mainly over 200 km, not one contained a Cuckoo's egg or chick. The three occasions on which I have watched Cuckoos go to lay in Meadow Pipits' nests were at altitudes of 120, 260 and 290 m.

Coulson (1956) estimated annual adult mortality at 57–58%, while Seel and Walton (1979) gave the similar figure of 54%. Meadow Pipits thus share with many small birds a short life expectancy, and a rapid turnover in population. They are the chief prey species of the Merlin, are commonly taken by Hen Harriers, and figure also in the diet of Kestrel, Peregrine, Buzzard, Raven and Short-eared Owl. Though usually well concealed, the nests are vulnerable to ground predators, notably Fox, Stoat and Weasel. They are placed among tussocky grasses, dwarf shrubs, cotton grass or rushes; and on flat ground, steep banks or even broken crag faces. Mostly they are found by accidentally flushing the tight-sitting female, which appears to do all the incubation.

As an insectivorous bird, the Meadow Pipit is greatly influenced by the seasonality of its food supply, and has to seek other winter quarters. Flock-

Fig. 20. Songbird of the hill grasslands: Wheatear on the alert.

ing often begins by mid-July, but the birds seem to roam the hills for some time, for there are usually frequent parties to be seen in most uplands well into August. Return movements to the low country vary according to the late summer weather, but it is usually mid-September before most of them have gone.

SKYLARKS are almost as widespread in the uplands, but usually far less numerous than the Pipits. They are especially birds of the hill grasslands, rather than heather and bilberry moors, but there are surprising numbers on some of the extensive peat moorlands of north-east Scotland. This is another bird with a fairly wide altitudinal range, and goes high on some of the sheep-walks, up to 900 m. Nesting is confined to flat or gently sloping ground, and the nests are typically more open than those of the Pipit, though sometimes a grass tussock gives partial concealment. The Skylarks return to the hills in February in open seasons, but do not reach the higher levels until later, and the end of summer sees their departure again.

The three chats are all quite typical hill birds. The WHEATEAR is the most widespread and numerous, though its numbers have fallen off in certain areas in recent years. This is perhaps the earliest of our summer visitors, and the first Wheatears used to appear on the Lakeland fells regularly around 21–23 March, but they now seem to be a week or more later on average. They are birds of grassland sheep-walks. The limestone hills of the Pennines are ideal terrain for them. Favourite nesting places are in crannies under large rocks, and in the crevices of block screes and drystone walls, but they will readily use vacant rabbit holes, as they do on the southern downs and brecks. Wheatears are frequent within the montane zone, and scattered pairs nest on many high block-strewn plateaux and in sheltered stony corries. Searches for Snow Buntings in these high places often see hopes raised and as instantly dashed as the flick of a small, pale passerine resolves itself into . . . only a Wheatear. High-altitude nests are often lined with Ptarmigan feathers and, in one instance, the ruddy breast plumage of a Dotterel was recognisable, the owner itself possibly having fallen prey to a Peregrine.

WHINCHATS are also summer visitors, but usually arrive about a month later than the Wheatears. They are widespread upland birds, but irregularly distributed and in greatly varying numbers, both geographically and from year to year. They are birds of the lower hill slopes, especially where there is long, tussocky grassland, bracken or heather, and their nests are usually well concealed in relatively thick vegetation. Newly afforested ground, in which the young trees provide song-posts, and the grassland or heath has grown rank and dense, suits them ideally. This bird also sometimes plays host to the Cuckoo. Its relative, the Stonechat, is most typically a bird of coastal and lowland heaths of heather and gorse, but is also widespread on similar ground (see p. 114) on the lower hill slopes.

Like the Whinchat, it also favours the pre-thicket stage of new afforested ground, and both species may be seen 'chacking' conspicuously from the tips of young conifers. Their habitats overlap considerably but Phillips (1970) found that, on a shared southern Scottish moorland, Stonechats preferred to feed in heather and Whinchats in grassier ground. Where the two bred close together, the Stonechat appeared to be the dominant species.

The RING OUZEL is another widespread and common bird of the uplands, though very sparse in Islay, the Outer Hebrides, Orkney and Shetland, and less numerous in Ireland. Otherwise, the map of its distribution almost defines the mountains and moorlands, from Bodmin Moor and Dartmoor to the north coast of Sutherland (Sharrock, 1976). It is one of the few species to nest regularly within the montane zone, though the bulk of the population lives at lower levels. The Ring Ouzel shares with the Wheatear the claim of being the earliest of our upland summer migrants to arrive, usually towards the end of March. From this time onwards, dawn chorus in the mountains often begins with the clear triple note of a cock Ring Ouzel carrying far across the hill.

While evidently preferring heathery hills, Ring Ouzels are typical birds of the sheep-walks, especially where there is much rocky ground. Nesting habitat varies widely, but one of the most favoured is the steep, rocky bank of a secluded little stream, preferably grown with long overhanging heather or other rank vegetation. Sheltered niches in crags, both on open slopes and in rocky gullies, or in old quarries, are often chosen. Drystone

Fig. 21. Ring Ouzel feeding young. The Mountain Blackbird is a summer visitor to this country but usually rears two broods in its upland haunts. Edmund Fellowes.

walls and abandoned cottages or old buildings frequently provide sites, and in the Pennines some nests are below ground level in the sides of limestone pot-holes. Juniper or other bushes are locally used, and also young trees forming the edge of a hill plantation (Poxton, 1987). Ring Ouzels sometimes nest on almost flat ground, amongst grass tussocks, or on the steep rim of eroding peat haggs. The nests are beautifully made and durable structures of tough hill grasses such as *Nardus*, and will last for several years in sheltered sites. It is not the easiest of nests to find, since the male's stances are often at some distance, but the sitter flies off with a noisy chatter of alarm when a person comes fairly close. When there are young the parents can be more easily watched carrying food, and they are then bold birds, returning to scold and flutter around close by if the nest is inspected.

Fig. 22. Prospective foster parents: Meadow Pipits and Cuckoo.

Ring Ouzels are often double-brooded and the female may settle to incubating a second clutch while the male takes charge of the fledglings of the first brood. Poxton (1986) found that although a Pentland group of Ring Ouzels were nesting on heather ground, they appeared to feed during the nesting period largely on earthworms, taken on the better pastures in the vicinity. Earthworms in the uplands are confined to the more fertile loam soils, but the bird nests quite widely in poor habitats where these are scarce or absent, so that other invertebrate food seems likely to be important in places. Breeding numbers appear to fluctuate from year to year, though Poxton (1986) found that in one valley of the Pentland Hills, Midlothian, numbers only varied between 5 and 7 pairs during 1979–84. An average spacing between pairs of 500 m in suitable habitat is quite a high density, and 1000 m probably more usual on sheep-walks. Pairs are often distributed more or less linearly along hill valleys or slopes, and it is not often that suitable nesting habitat extends continuously over large areas. In late summer, Ring Ouzels turn to the berries of the moorland shrubs and rowans. Family parties or, sometimes, larger groups move around the hills, evidently in search of this vegetable food.

Ring Ouzels are reported to have declined in some districts, and over a long period (Sharrock, 1976; Thom, 1986). There have been recent local decreases in the northern Pennines, but this is a bird for which there is a dearth of precise, long-term information, and it is clearly still a widespread species over much of our hill country. One local cause of permanent decline is clear, however: the blanket afforestation which has obliterated so much of the Ring Ouzel's open moorland habitat in Wales, the Cheviots, Southern Uplands and parts of the Highlands. We shall consider this later.

The CUCKOO is a familiar bird voice in the uplands though it mostly frequents the fell-foot country and the lower slopes, and does not follow its main foster parent, the Meadow Pipit, far up the mountains. As an upland species its biology is not at all well known. Numbers seem to vary from year to year, but whether in relation to those of Meadow Pipits is unclear. Other host species in the uplands are Whinchat, Skylark and Pied Wagtail. Bolam (1913) noted that on railway embankments near Bala, in Wales, Cuckoos habitually laid in the nests of Meadow Pipits but never once in the far more numerous nests of Tree Pipits. Yet in some areas, Tree Pipits are evidently fairly frequent Cuckoo hosts. In Merlin haunts, Cuckoos in flight often raise hopes, but then realisation dawns that this is only an impersonator. The overlap in habitat is risky, for Cuckoos and their young not uncommonly appear amongst the remains on Merlins' feathering blocks. They are also a quite frequent item in Peregrine prey, especially in the Highlands.

The PIED WAGTAIL is perhaps a doubtful candidate for inclusion in our list. It is most often around the hill farms, nesting in outbuildings and drystone walls, but also in places in the lower crags and along rocky stream sides, there overlapping with Grey Wagtails. Of the familiar lowland passerines, the WREN is quite convincingly a hill bird. With a species so much associated with gardens, farmlands and woods, it may come as a surprise to hear the cheerful song or discover the nest high in some bleak corrie of the mountains. As usual most of the nests found casually are the unlined 'cock' nests, used by the males as roosts. They are usually in steep banks or crag faces: sheltered rocky glens are favourite haunts, and one nest was almost behind a small cascade in a Dipper-type site. The highest elevation at which I have seen a nest with eggs or young was 620 m, but breeding probably often occurs much higher. Among cock nests in noteworthy situations was one at 830 m in a cave-like recess in the great escarpment under Snowdon summit, and another at about 900 m on the cliffs of Caenlochan Glen in Angus, both habitats being important localities for arctic–alpine plants.

Fig. 23. A resident of the rocks: Wren nest building.

One intriguing feature is that the nests chosen for laying are frequently lined with feathers collected from Peregrine kills. Most craggy haunts of Peregrine and Raven have their Wrens, and one cock nest was only 3 m below a Golden Eagle eyrie, thereby maintaining the legendary association between this feathered mite and the King of Birds. The hill-going Wrens feed especially among long vegetation, such as heather and bracken, and within deep rock crannies in screes and stone walls. They tend to seek this cover readily and make short flights between hiding places. Perhaps these retiring ways help to make the Wren a little-known upland bird, for few writers have much to say about it.

Waders

The last bird group of the sheep-walks is the waders. Several species are especially characteristic of the grasslands, though they frequent other kinds of uplands as well. The LAPWING is par excellence the bird of the more fertile grasslands, both on the enclosed land of the upper farms, and the unenclosed hill extending upwards to at least 600 m. As an earthworm feeder, in particular, it needs good loamy soils, and its presence is often associated with that of moles. The Pennines have a large extent of limestone pasture, and must be one of the most important districts in Britain for nesting Lapwings, now that the species has declined so widely in the intensively farmed lowlands. During open winters the Lapwings return to the uplands in February, and their wheeling, tumbling flight as the birds engage in territorial and courtship display is one of the first signs of approaching spring. Breeding density on limestone ground is often high, reaching 60 pairs/km^2 locally, so that territories here are quite small.

The 'greens' or improved pastures around the last hill cottages and crofts – some of them now deserted and ruinous – typically have their pair or two of Lapwings. The bird now and then takes to nesting on the acidic heather moorlands or even the blanket bogs, and the high tops of the Pennines have sometimes drawn a few pairs to try their luck. I have seen a nest within a short distance of the summit cairn of Crossfell, at nearly 900 m, but these high nestings usually have little success and do not become a regular habit. If the weather remains mild, many Lapwings begin to lay during the first half of April. Their open, unconcealed nests, often consisting of a substantial pad of grass, would seem vulnerable, and many are lost to foxes, crows and gulls. The Lapwing is a noisy, aggressive bird in defence of its nest, and winged predators are sometimes successfully driven off before they can find the eggs. Yet, despite its aerial agility, the bird often falls victim to the Peregrine's onslaught. Foxes sometimes catch incubating birds unawares at night. So there is usually some failure of first nests, and repeat layings may be found until well into June, as unlucky pairs have to try a second or even a third time. When the chicks attain flight, after 5 weeks, the families begin to flock and move away, and some have already forsaken the nesting grounds by late June.

Although Golden Plover often share feeding places with Lapwings on the enclosed hill pastures, their nesting haunts usually begin where the Lapwing ground gives out: on the acidic grasslands, heather moor and blanket bog beyond. Yet in the Pennines, both species nest side by side on plateau limestone grasslands, and the Golden Plover attains a particularly high breeding density on some of these habitats. On the whole, though, this bird is less numerous on the sheep-walks than on the grouse moors.

The COMMON SNIPE is an extremely widespread bird in the uplands, but it remains one of the least studied species. It would be unwise to attempt even a local estimate of numbers, but my impression is that they vary from year to year. Birds are commonly flushed from their feeding places in little marshes and other wet ground, usually where there is enrichment from lateral water flow, and from the lowest levels to well into the montane zone. Drumming Snipe are frequently to be heard, and their sound comes eerily through the dusk or moonlit nights. The number of nesters remains mysterious: in the course of my hill wanderings, I have

Fig. 24. Curlew about to incubate. The extent of their nesting habitat has decreased greatly in heavily afforested districts. R.H. Fisher.

come across only 10 nests, two of them at 800 m. They appear to be most numerous on the richer soils, but perhaps the best habitats are the damp, rushy and only half-drained or improved pastures of the upland valleys and hill bottomland. In some areas, the frequency of their remains in Peregrine kills may give a general indication of Snipe abundance. The practice of draining wet hill pastures and moorland is particularly unfavourable to this bird, by causing a deterioration or elimination of its preferred feeding places.

Another wader of the marginal land with its rushy pastures and wet meadows is the REDSHANK. This bird is perhaps less uniformly distributed and numerous than the Snipe, but is easier to count, especially from its noisy and conspicuous behaviour when the young have hatched. It ranges upwards to the limits of the submontane zone, but is better represented on the grouse moors than the sheep-walks. On the open hill, it is perhaps nowhere more numerous than in the Pennines, but even there is somewhat patchy in occurrence, with scattered single pairs at fairly wide intervals.

The last but probably most characteristic wader of the sheep hills is the CURLEW. It occurs through a wide range of lowland country as well, but whereas the Redshank has tended to move upwards into the hills during this century, the Curlew has spread downwards. Its rippling song and graceful display flight are amongst the most evocative of appeals to the senses as springtime reaches the moors (Fig. 2). On many of the wide sweeps of *Molinia* ground in the Galloway uplands the air could seem alive with their territorial activity on fine days in April. Nowhere was the bird more numerous, though the highest numbers were on the sedimentary rocks and it was much sparser on the granite. In general, the Curlew varies greatly in breeding density between different areas, and has shown marked changes in abundance in some places.

Philipson (1954) knew the Curlew as numerous and widespread over the entire uplands of west Northumberland and north-east Cumberland around 1912, but observed a steady decline over the next 30 years. At high density, he found each pair to occupy an area of 8 ha with these territories vigorously defended by the males against rivals. Most Curlews nest within the submontane zone, below 600 m, but in the Pennines some go higher and a pair regularly nested on the summit of Knock Fell at nearly 800 m. Curlews are often numerous breeders on the enclosed pastures of the marginal lands, and the upland hay meadows.

Although Curlews are aggressive in the defence of their nests against predators, often hustling away intruding gulls and Crows and scolding noisily but warily any discovered Foxes, their eggs and young still fall frequent prey. Remains of sucked eggs may explain many incomplete incubated clutches, and a proportion of pairs in any area usually has repeat clutches

at a time when the rest have hatched. At one time, hill shepherds often rel-
ished their large eggs for the table. The nests are not the easiest to find,
however, and the sitting birds can show considerable skill in leaving their
eggs without giving away the location. The staccato triple alarm note
infallibly tells of a nest nearby, but when the young have hatched, parental
concern is expressed in continuous agonised shrieks, as the Curlews wheel
low round the human disturber. One is forcibly reminded of how the
resentful Curlews and Lapwings are supposed to have betrayed the old
Covenanters as they sought refuge from the persecuting Dragoons on
those wide, open moorlands of the Southern Uplands.

With periods of 70 days between laying and fledging, it is usually early
July before the first young Curlews can fly, but by the middle of the month,
there is a steady exodus from the uplands to the lower ground, and then
the coasts. The birds appear to leave mainly in small parties or even
families, rather than as large flocks. During winter, our breeding stock of
Curlews becomes mixed with passage birds and immigrants from farther
north in Europe, making the species one of our most numerous waders.
This may tend to disguise the steady decline of the native population, for
in both Britain and Ireland, this is a species which has lost a great deal of
breeding habitat through afforestation, and stands to lose considerably
more.

The extensive post-1940 agricultural improvement of the damp pastures
and meadows of the enclosed marginal lands has caused a tremendous
reduction in good habitat for these grassland waders. In 1985–87 Baines
(1988) compared their breeding densities on improved and unimproved
enclosed hill grasslands in the northern Pennines. Improvement by draining, fertilising and sometimes ploughing and re-seeding resulted in virtual
disappearance of Snipe, and marked reduction in densities, and proportion of fields used by Redshank (81% loss), Curlew (82%) and Lapwing
(74%). Loss of cover, in the previously numerous rush clumps, as well as
deterioration in food value, was important for the first three species; but
reasons for decline of the Lapwing were less clear. There was no evidence
that the birds redistributed themselves after reclamation, and the declines
seem likely to be real population reductions. Given the scale of marginal
land reclamation in the past 50 years, and the fact that Baines found combined wader densities of up to 140 pairs/km^2 on the best unimproved
grasslands, the overall losses nationally on these habitats have been serious. Only the Oystercatcher showed an increase on improved pastures,
but the numbers involved were small.

An associated effect is that when grazing animals reach a high density
on enclosed pastures, there is a strong risk of them trampling the eggs or
chicks of ground nesters such as Lapwing, Snipe or Redshank (Green,
1986). The effect on breeding success can be serious and it may even discourage some birds from nesting.

Post-1940 reclamation and enclosure of the *unenclosed* hill ground have also had a considerable impact on the upland bird fauna, as have improvement and increased sheep stocking of the open moor. The cutting of extensive systems of drains in wet moorland ('moor-gripping') has dried out many of the botanically interesting flushes and little marshes which are favourite feeding places of Snipe and probably other waders. Its effects on birds, including the possible hazards of the drains to the chicks of Grouse and waders have not been studied. Afforestation has, however, been the most profound of recent land use changes in its effects. Since this activity affects also the heather moorlands, it will be dealt with in the next chapter.

Birds of streams and lakes

The upland streams, whether in sheep-walks, grouse moor or deer forest, have several typical birds best dealt with as a group. Though absent from the Isle of Man, parts of the Hebrides, Orkney and Shetland, the DIPPER is a familiar stream-dweller over most of the uplands, ranging from the main valley bottom rivers to the high cascading rivulets of the montane zone. It is an engaging and conspicuous little bird, bobbing about on the stones of its home stream, to which it clings through the hardest of winter weather.

Fig. 25. The King of the Waterfall: Dipper on a cascading stream.

Shaw (1978) points out that this fidelity reflects the build-up of invertebrate biomass during winter, to reach a maximum in early spring. Dippers are early breeders and their song and courtship activity often enliven the hill streams in the depths of winter. Many of the lower-altitude nests have eggs before the end of March.

The vicinity of cascades and waterfalls are favourite nesting haunts, and in Norway the bird is known as the King of the Waterfall. Nests are sometimes placed in caves behind falls, and the great majority are directly above or even overhanging the water, into which the young take a header on leaving for the first time. Any steep rocks above a pool, or bridges and other constructions with suitable lodgement may hold a nest. Where these are lacking, tree roots, fallen trunks, large boulders in the stream bed, and banks with recesses or long vegetation hanging over the water may be used. The nest of moss is often well concealed or inconspicuous, with the contents dry and well protected under the overhanging dome. Shaw records that the same nest is typically used for successive clutches within the same year, but many pairs have from one to several alternative nest sites used over a period of years.

Since populations on any river system can be accurately counted, the Dipper has been a favourite subject for study, and a good deal is known about its breeding biology in Britain. Shaw (1978) found that 87% of the BTO nest records were for sites below 300 m; although this sample may not be nationally representative, the bird does appear to be more numerous at the lower elevations. Estimates of breeding density and territory size are, however, made difficult by the variations along any one stream in availability of nesting sites and suitability for feeding. Linear spacing of occupied nests is an appropriate measure, but as Shooter (1970) points out, the width of the stream is important to the amount of feeding ground: he estimated that in good habitat, a pair of Dippers needs about 0.4 ha of stream area, which in his Derbyshire rivers was equivalent to one territory every 540 m in linear distance. Shooter found that Dippers prefer shallow water, and that the much greater food potential of certain limestone rivers was nullified by their deep water, so that on both limestone and gritstone streams density was similar at about 1 territory per 1.6 km. On the River Eden in Westmorland, Robson (1956) found that 23 territories averaged 430 m long, with similar densities on both limestone and sandstone for occupied river sections, but a greater overall suitability for sandstone reaches. Scottish breeding densities in Midlothian and Angus averaged around 1.4 km per pair.

The Dipper was once one of the most constant upland birds within its habitat, but in certain districts it has declined of late. Ormerod *et al.* (1985) connected the disappearance of Dippers from headstreams of the River Irfon in South Wales with extensive afforestation of the catchments, and associated acidification of the stream waters, with reduction in the bird's invertebrate food. Dippers were still present on unafforested streams in the same area. In a later study over a much wider area of upland Wales, Ormerod *et al.* (1986) found that breeding Dippers were present on only 21 out of 74 sites (28.3%). Streams without Dippers had higher concentrations of filterable aluminium, lower mean pH, lower abundance of mayflies and caddis flies, and were on narrower rivers with fewer broadleaved trees on their banks and more conifer afforestation on their catchments, than those where the bird was present.

Dippers live especially on larvae of mayflies, caddis flies, blackflies, stoneflies and small fish, which Ormerod and his colleagues have found to show marked biomass decline in acidified streams, especially those within heavily afforested catchments. This pattern is consistent with the acidifying effects of atmospheric pollution as the underlying cause, but accompanied also by the influence of conifer plantations in enhancing drainage water acidity through the 'scrubbing' action of their foliage. Juliet Vickery (personal communication) has also found a similar reduction in Dipper numbers on acidified streams of the Galloway granite uplands.

Dippers seldom leave the water's edge, and their frequent flights usually hold faithfully to the stream course. During July and August moult leaves the adults with little power of flight, and they tend to become silent and skulking then. Birds living by or close to lakes sometimes show the curious habit of flying out and alighting on the open water, well out from the shore, there to swim around and dive 'like little ducks' (Bolam, 1913). Although a hardy bird, able to feed even under ice at times, the Dipper nevertheless inevitably suffers during the hardest winters, when prolonged frost gradually seals over all its feeding places. Yet, with often 2 broods in the year, and these usually of 3–4 young, it seems to recover in numbers within a few years of quite severe declines. It is a resilient species under such natural adversities, but less secure against the pervasive interferences of human beings in its environment.

The GREY WAGTAIL is another characteristic upland stream bird, with a wider distribution than the Dipper on lowland rivers, but showing similar absences from some of the Scottish islands. In some uplands it is less numerous than the Dipper, but ranges and breeds up to a similar elevation. The favourite habitats are stream ravines and waterfall glens, where the nest is usually placed in the steep, rocky sides, but may also be in rocks some distance from the water. Not uncommonly it is built on top of an old Dipper nest. When there are eggs, the delightful owners are not inclined to reveal its location, and will shift repeatedly from one perch to another, often over quite a wide area, so that the watcher becomes baffled and gives up.

The Grey Wagtail has been little studied in the uplands. There is no evidence that it has been adversely affected by stream acidification, in the same way as the Dipper, perhaps because its food is less exclusively aquatic (Tyler, 1987). It is, however, another species to suffer badly during severe winters, but seems to recover rapidly, perhaps because it is partly migratory, and relatively widespread in the warmer lowlands, anyway (Sharrock, 1976).

The COMMON SANDPIPER is a characteristic bird of streams and lake margins over most of upland Britain and Ireland. A summer visitor, it is one of the later species to nest, though full clutches are usual from around mid-May at lower levels. Most pairs nest within the sub-montane zone, but Red Tarn on Helvellyn (720 m) is a breeding haunt, and in the Cairngorms Sandpipers nest regularly in arctic settings by elevated lochs, occasionally as high as 1100 m. Confined ravines and steeply descending cascade streams are avoided, and rivulets or some small tarns are unattractive to Sandpipers, but many moorland rivers and shores of large lakes have long stretches of suitable habitat, where pairs tend to space out along fairly even lengths of territory. My impression is that on some streams of the Scottish Borders, at least, numbers (and breeding density) can vary markedly between different years. Studies elsewhere suggest more constant breeding populations.

The Nethersole-Thompsons (1986) found an average linear breeding density of 1–2 pairs/km of river length in north-west Sutherland, but on Speyside there were about 6 pairs/km around Loch Morlich and on the River Dorback.

On Peak District rivers, Yalden (1986) found that 18 territories averaged 244 m on occupied sections, giving a density of 4.7 pairs/km over prime stretches of river. Spacing between pairs is often much wider on both upland and lowland rivers, at 0.2–1.0 pairs/km. Some types of river and lake are clearly more suitable than others. Yalden (1986) found that the bird prefers wider, more shingly stretches of river. Many pairs used both rough and improved grassland (including enclosed fields and hay meadows) within 100 m of the river for feeding, especially in the first fortnight after arrival. Here, they showed little territorial interaction, in contrast to their behaviour along the riverside. Very young chicks fed especially in short grass, but at over 5 days old they spent most of their time, including feeding, amongst the boulders and shingle of the river bed.

The nest sites vary a good deal. Some are close to the water's edge, and may be within the flood zone, but more typically are farther back and above extreme high water line. They are usually where vegetation provides some concealment. A favourite situation is a fairly steep moraine bank, where a small, sheltered hollow is used. Where roads or railways run close to a lake or main river, their sides and banks are often used. Some nests are well back from the water's edge – up to 300 m or even more – and some are among open tree growth. This bird makes a more substantial nest than any other of our waders except perhaps the Lapwing: a neat cup of dried grass.

Sharrock (1976) reports long-term declines from various parts of Britain and Ireland. The Common Sandpiper has withdrawn from the uplands of south-west England, and is sparse in some of the hill country in southern and eastern Ireland, but it has also declined in Orkney and eastern Scotland. It does not appear to have been affected by stream acidification in the same way as the Dipper, but its response to afforestation deserves study. Sandpipers continue to nest along at least the larger streams within new conifer forests, where the planting line is held well back from the water (as is usual practice nowadays), but whether in the same numbers as before afforestation is not known. The bird is the fisherman's constant early summer companion, and there are worries that excessive disturbance through fishing may be causing local decline, e.g. along the shore of L. Ken in Galloway (A.D. Watson).

Bolam (1913) believed that Sandpipers sometimes carry their chicks, in the manner of a Woodcock, but such behaviour seems not to have been noticed by others. This is, however, a somewhat neglected species, deserving closer study. Its stay in this country is fairly short, and by early July a good many pairs have already left the uplands.

Although they are predominantly coastal and/or lowland, both the OYSTERCATCHER and the RINGED PLOVER have some claim to be considered as members of the upland breeding bird fauna. Both nest on the shingly, gravelly and sandy margins of streams and lakes in northern England and Scotland, mostly below 370 m in the main valleys amongst the hills. The Oystercatcher is the more numerous and widespread of the two among the uplands, and uses a wider range of habitats away from water, including enclosed fields, moorland and even bogs. It has increased greatly as an upland bird in northern Britain during this century, and some of the riverside breeding densities on the Spey compare well with those in good coastal habitats (Nethersole-Thompson and Nethersole-Thompson, 1986). Ringed Plovers in Britain seem not to resort to high montane fell-fields which are equivalent to the species' breeding haunts on Arctic barren grounds, where a separate ecological adaptation appears to have evolved. Scattered pairs nest by the sandy shores of lochs amongst the bogs of the Flow Country, and beside other mountain lakes. Inland nesting groups on more productive lochs and rivers seldom reach the densities found in many coastal haunts. In Galloway, the breeding strength of Ringed Plovers at Clatteringshaws Loch varies according to water level and extent of exposed shingle around this hydro-electric reservoir (A.D. Watson).

The MALLARD is an extremely widespread duck in the hill country, though mainly in scattered pairs along the upland valleys and on the lochs, in the breeding season. The birds often frequent wet places away from the rivers, and their nests are in a variety of situations, such as rocky stream banks, tall heather, bracken, rushes or long grass. Many nest sites are far from water, for greater safety. Hill Mallard are shy birds and rather little is known about them. They form into parties of varying size outside the breeding season, but these are often of only a few birds, and bigger flocks tend to be on little disturbed tarns or larger lakes.

The two handsome sawbill ducks, GOOSANDER and RED-BREASTED MERGANSER, are northern rather than upland, but are quite characteristic of the bigger rivers and lakes in the hill country, especially in Scotland. Though still a rarity in Ireland, the Goosander is the more widespread hill species, and the Merganser shows a distinct preference for coastal moorlands, feeding especially in tidal waters (Sharrock, 1976). For nest sites, Goosanders favour tree holes in open woodland or scattered trees near water, but will also nest among blocks, tree roots or crannies in rocky glens. Mergansers typically lay in long heather or other rank vegetation, often on islands in lakes and rivers. Both species are subject to a great deal of legal persecution because of their predation on young salmon, which form a significant part of their diet (Mills, 1962). Both have spread a good deal in recent decades, even so, and still appear to be extending their range southwards (Sharrock, 1976).

Buildings as bird habitats

Man has left his most obvious imprint on the uplands in the form of buildings, both used and disused. Shepherds' cottages with their outbuildings are still widely occupied, but many in the remoter places have been abandoned and are in various stages of disintegration. The summer dwellings of sheep and cattle men in the former days of transhumance are now mostly ruinous: the hafod in Wales and the shieling in Scotland. In the Pennines, the lower hillsides have characteristic stone barns ('field houses') solidly built of gritstone blocks and used for the storing of hay. Old mine and quarry buildings in various stages of decay are a local feature, and along disused railways are viaducts and bridges (though often subject to demolition). All these structures are quite a noteworthy habitat for nesting birds, including both upland species and others which have not been thus regarded.

At the lower levels, old buildings commonly provide nest sites for Jackdaws, Starlings and Stock Doves. Swallows locally occupy barns and derelict buildings in the hills and a nest was found at about 700 m in an old Pennine mine hut. House Martins here and there nest under the eaves of upland houses or on more imposing structures such as reservoir dams. Although Swifts so often hawk insects over the high tops, nesting in natural rock faces has yet to be proved, and the birds mostly come from the hill country villages, though they occasionally nest in old railway viaducts on the moors.

Some of the predators use old buildings for nesting. Ruinous dwellings amongst the hills have become increasingly important for Barn Owls as the bird has declined in the lowlands, and Kestrels sometimes use similar sites, or holes and ledges on viaducts. Hill Ravens have nested on a variety of structures: deserted cottage roofs, mine buildings and pit-head gear, electricity pylons and disused railway viaducts. At least one pair of Peregrines has established a novel nesting place on an old viaduct. Choughs have quite often nested in abandoned buildings and do so regularly on the Isle of Man and Islay. Ring Ouzels frequently resort to buildings for nest sites, and a variety of passerines can be found using them in one upland locality or another, including Grey Wagtail, Pied Wagtail, Wren, Redstart, Pied Flycatcher and Tree Sparrow. Drystone walls provide nest sites for Wheatears, Ring Ouzels, Pied Wagtails and Wrens.

3

The grouse moors

Certain hill areas have long been managed primarily as sporting preserves for the Red Grouse. Most grouse moors have in common a gently contoured topography and a prevalence of ling heather, *Calluna vulgaris*, the bird's staple food. Gentle relief is preferred because it is easier to shoot Grouse there than on steep hills, especially by driving them over the guns. Grouse moors have a somewhat eastern distribution, partly through the prevalence of much steep and unsuitable terrain in the west, but mainly because of climatic differences. Heather grows well enough in the wetter west, but is typically accompanied by vigorous grasses and their allies which have a distinct competitive advantage under burning and heavy grazing. Heather-dominated communities are thus less extensive and

Fig. 26. Moorlands of Strath Don, Aberdeenshire. Classic dry heather grouse moor habitat, showing rotational burning pattern with heather patches of different age. Typical breeding ground of Golden Plover, Hen Harrier and Merlin.

more difficult to maintain in the west, and have retreated here over the last few centuries. Grouse moors nevertheless still cover around 1.5 mha of Britain (Hudson and Watson, 1985).

Dry, heathery grouse moor is exemplified by the North York Moors, the Lammermuirs, and the hills of Glen Dye and Strath Don. Even in fairly dry areas, gentle relief often leads to waterlogging and peat formation, so that many grouse moors have extensive bog, especially on plateau watersheds. The highest stocks of birds have, in fact, consistently been on moors with both dry heather ground and blanket bog, where heather shares dominance with cotton-grass. Some lowland raised bogs and upland flows have been managed as grouse moor, and here dry heather ground is virtually absent.

Most birds of the sheep-walks are also found on the grouse moors, but this chapter will concentrate on species especially characteristic of these heathery uplands. While possessing many of the typical grouse moor species, the bird fauna of the flows is sufficiently distinctive to merit a separate treatment, in Chapter 5.

Formerly cherished as our only endemic bird species, the RED GROUSE of Britain and Ireland is now demoted to a sub-species of the widespread circumpolar Willow Grouse. The original habitat of the Red Grouse was evidently open pine and birch forest with much heather ground, treeless peat bog, and the zone of heather moorland above the tree line. Denser

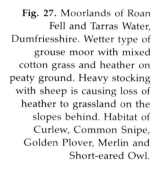

Fig. 27. Moorlands of Roan Fell and Tarras Water, Dumfriesshire. Wetter type of grouse moor with mixed cotton grass and heather on peaty ground. Heavy stocking with sheep is causing loss of heather to grassland on the slopes behind. Habitat of Curlew, Common Snipe, Golden Plover, Merlin and Short-eared Owl.

woodland was unsuitable and there the species was replaced by Black Grouse and Capercaillie. The Red Grouse has thus been particularly favoured by the huge expanses of largely treeless dwarf shrub heath created through human forest clearance over the past few thousand years.

The Red Grouse is a hardy fowl, seldom leaving its moorland haunts, though where cereals are grown on the upper farms, some birds glean in the fields after harvest. After heavy snow, Grouse often assemble in packs and move downhill to forage about the lower ground until conditions improve. They can also burrow into the snow, and live in their tunnels, which give shelter and access to buried food. With snow-melt the Grouse become territorial again, and during still, fine days at the end of winter the moor resounds with their 'becking' as the cocks display and confront each other in vigorous challenge (Fig. 31).

Although the Red Grouse probably occurred during the colder early Boreal period on heathery heaths of southern and eastern England, it is now restricted to upland and northern habitats, and probably characterises these better than any other bird (see Fig. 29). The southernmost populations, on Dartmoor and Exmoor, were reintroduced and are above 400 m, but in west Wales, the Morecambe Bay area and the Solway, Grouse occur on 'peat mosses' close to sea level. They are generally distributed at low levels on coastal moorlands and lowland heaths or bogs throughout the Highlands and Islands. On the mountains they occur commonly up to the limits of good heather ground at 800 m, and there overlap the lower limits of the Ptarmigan zone. In Ireland, they are widely distributed, occurring in most hill areas, and on the raised bogs of the Central Plain, but are nowadays seldom in high numbers anywhere. Breeding densities of 80 pairs/km^2 are now regarded as high; on the poorest ground there may be only 1–3 pairs/km^2.

Fig. 28. World breeding distribution of Willow Grouse (from Voous, 1960).

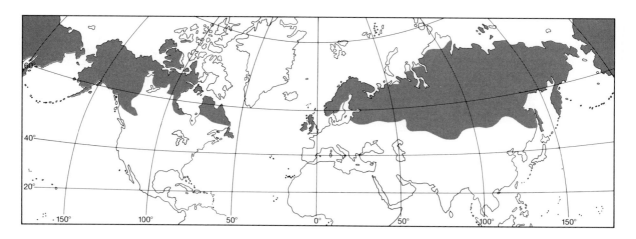

Although there are few hill districts without Grouse, they are now scarce on many grassy sheep-walks, and occur only where areas of heather survive. They are very local and sparse in west Wales, Lakeland and parts of the Southern Uplands, and at low density through much of the western Highlands and Islands, and in western Ireland. Good Grouse numbers occurred in Glen Almond, Perthshire, in the early 1960s, on grassland with little heather but much bilberry. The economic importance of the Grouse and its varying fortunes have caused it to become one of the most studied

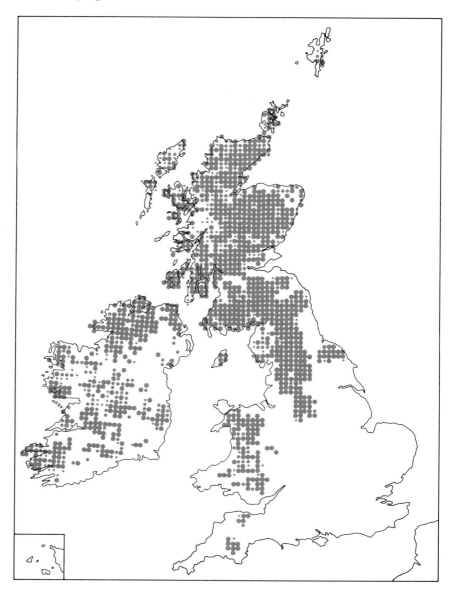

Fig. 29. Breeding distribution of Red Grouse in Britain and Ireland during 1968–72 (from Sharrock, 1976).

birds in the World. After a little background on Grouse shooting, I shall attempt to summarise present knowledge of its population dynamics reported in a now extensive literature.

Grouse shooting was well established by the late eighteenth century, but most grouse moors came into being around 1800–1840, and the sport reached its peak after the 1860s, when the muzzle-loading shotgun was replaced by the more efficient breech-loader. The hey-day of Grouse shooting was probably the following half century. Yet it remains a major British field sport, and important to the economy and ecology of many upland areas. Grouse shooting became popular because appropriate management was able to increase stocking density of the bird to phenomenal levels, far beyond those in natural populations. The particular management to boost numbers rested on special permanent custodians of the moors, the gamekeepers, whose job it was to burn the heather on a carefully planned rotation, place grit and water for the birds if necessary, and eradicate all 'vermin' suspected of predation. The number of birds shot, especially on the opening day of the season, was the measure of success, and the 'Glorious Twelfth' became a key event in the social calendar.

When Gladstone (1930) compiled his list of record bags, the largest number of Grouse shot in one day on one shoot was 2929 by 8 guns on 12 August 1915 on a beat of the Bowland Fells in the western Pennines. On two adjoining beats, 1763 and 1279 birds were killed on the 13th and 14th, making a total of 5971 Grouse for three consecutive days. The bag for the season was 17 078 Grouse, from about 17 000 acres. Broomhead Moor in the Peak District was a noted shoot where single day bags of 2648, 2743 and 2843 Grouse were made in 1893, 1904 and 1913; and a total of 5000 birds from 4000 acres in one season. In Scotland, the record was 2523

Fig. 30. The 'glorious Twelfth': driven Grouse over the butts.

Grouse killed by 8 guns on 30 August 1911 at Roan Fell, in the Southern Uplands. Another 1266 birds were shot here 8 days later. The record for Wales was on Ruabon Mountain, Denbighshire, where 1774 Grouse were shot by 8 guns on 13 August 1912. In Ireland, the highest tally appears to be 358 birds to 13 guns at Powerscourt, in the Wicklow Mountains, on 25 September 1890.

Population ecology of Red Grouse

Despite all the success in late Victorian Grouse management, worries were expressed about 'epidemics of mortality' on many moors. Although numbers usually recovered again after a few years, there was often severe disruption of the sport and financial loss. These population crashes were so economically damaging locally that in 1905 a Committee of Inquiry on Grouse Disease was set up by private subscription to investigate the causes and advise on prevention. From the catastrophic nature of the declines and numbers of birds found dead, it was assumed that rampant disease must be involved. In 1911 the Committee published a lengthy report, *The Grouse in health and in disease*, the work largely of the Chairman, Lord Lovat, the Secretary A.S. Leslie, E.A. Wilson (of the Antarctic) and A.E. Shipley. A popular edition followed (Leslie, 1912). It was the first intensive ecological study of a bird species ever conducted, and one still well worth reading, as an example of an early scientific approach.

The main conclusion was that the chief direct cause of excessive adult mortality (i.e. 'Grouse Disease') was caecal infestation by the nematode threadworm *Trichostrongylus tenuis*, while the protozoan infection coccidiosis affected only chicks and was a secondary problem. Since the nematode was found in nearly all the 2000 Grouse examined, including many in apparent good health, its effect was thought to be conditioned by other factors which varied the level of individual infestation by the parasite, and also altered the birds' resistance to infestation. It was supposed that, because *Trichostrongylus* did not multiply in the body of its host, density of Grouse numbers appreciably affected exposure of individuals to ingestion of threadworm larvae.

The weight of a Grouse was regarded as an indicator of its health, and this was understood to depend especially on food supply. Analysis of crop contents showed some seasonal variation in diet, but the shoots, flowers and seed-heads of *Calluna* formed a yearly 81% of the food. Bilberry was important and over 20 plant species were commonly taken, including most moorland dwarf shrubs and their berries or seed-heads, and various herbs. By contrast, the chicks ate mainly insects (especially Diptera) and other invertebrates during their first two weeks, and then increasingly the buds, young leaves and flowers of various shrubs and herbs.

The Committee concluded that the amount of good heather, particularly during winter and spring, was crucial to Grouse health and numbers. Quality of heather and overall carrying capacity were thought to be optimised by rotational burning of one tenth to one fifteenth of the moor annually, to produce plenty of dense, young growth at any one time. A mosaic burning pattern of small patches of different age was best. The implications for Grouse numbers of the generally better condition of heather moor in the drier east, compared with wet western districts of Scotland, were also noted, as were the deleterious effects of prolonged drought, cold east winds and severe frost without snow cover. Food depletion through attack by heather beetle was viewed as a local and periodic problem.

The indirect effect of weather on food supply was considered to be enormous, but the direct effect on the birds slight. Even bad weather during the breeding season seldom had adverse effects on stocks for 12 August, though excessively wet and cold conditions could cause heavy chick mortality. Dampness favoured survival of *Trichostrongylus* larvae, but dry warmth promoted *Coccidium* infection. Dry conditions were nevertheless generally favourable to the Grouse.

The Committee clearly understood that an interaction of factors was involved in the population crashes. Adverse weather or increased population density could reduce winter food supply per bird and, hence, the resistance of the stock overall. When high density of Grouse increased threadworm infestation, conditions were ripe for heavy mortality to follow: hence the tendency for record Grouse years to be followed by outbreaks of disease. Most disease mortality was in early spring, when the rigours of winter and shortage of food most reduced the birds' resistance. The parasite could cause sterility in females or small clutches, but poor breeding success was attributed more to coccidiosis. Adult mortality in late winter and spring was regarded as more important than breeding performance in determining late summer stock size.

The solution to the 'disease' problem was thus seen as increasing the resistance of Grouse and reducing parasite infection by careful moor management. The Inquiry recognised that epidemic mortality had occurred right through the nineteenth century, and that few areas had escaped. Indeed, the map of outbreaks of 'Grouse Disease' from 1872 to 1909 (Leslie, 1912, p. 318) is virtually a map of grouse moor distribution. Outbreaks also had a somewhat cyclical character, in some areas with a 7 year recurrence, but elsewhere varying irregularly between 3 and 8 years. A latent propensity to epidemic death was regarded as an unfortunate constitutional characteristic of the species. While exacerbated by the high densities achieved through management, the problem could also evidently be mitigated by carefully adjusting management to local conditions.

During 1920–40, animal population ecology developed and showed that in some mammals and birds, regular cycles of increase followed by sudden decline were normal features. They appeared to occur especially in predators dependent on only one or a very few prey species (eg Snowy Owls, and Snow-shoe Hares and Lemmings in the Arctic; Short-eared Owls and Field Voles in upland Britain), but also in many game birds (Lack, 1954). From 1945, when interest in Grouse shooting was resumed, concern grew over long-term downward trends in numbers on some moors. Allowing for normal ups and downs, peak populations were failing in many places to reach levels commonplace even during the 1930s. The best moors had once produced yearly bags of 230–280 grouse/km^2, and bags of 100 birds/km^2 were commonly exceeded on many moors in England, Wales and Scotland. In certain districts, decline in shooting bags had been noticeable since 1900, but there was wide variation between different districts, and some showed a marked recovery again in the 1930s. Nowadays few moors anywhere produce bags exceeding 100 birds/km^2 (Hudson, 1986a).

The upshot was that a new study of Red Grouse ecology and reasons for population fluctuations was set up in 1956 in north-east Scotland. The work was begun by David Jenkins, but soon developed into a team effort under Adam Watson, Gordon Miller and Robert Moss, within the Nature Conservancy's research programme. This Red Grouse study is still in progress. I have tried to summarise its main findings, though there is space to acknowledge but few of the many relevant papers.

Effort focused first on population behaviour through the annual cycle, over one main study area at Kerloch Moor, Kincardineshire. Survival of breeding Grouse through the summer was usually good, and overall numbers on 12 August depended more on production of young, which varied annually. The combined numbers were then thinned by shooting. In autumn there was a further reduction in population through emigration of some birds. Those left on the moor during winter were paired and strictly territorial. The Grouse which quit had been displaced by the territory holders: they moved into more marginal habitats and appeared there to incur higher predation pressure and other hazards, so that many died. At the onset of the next breeding season, territorial activity became heightened, and there was a second displacement of surplus Grouse to other ground, again with enhanced mortality (Jenkins, Watson and Miller, 1963).

The Grouse thus evidently adjust their own numbers through the year, by territorialism within a peck-order, leading to higher mortality amongst low rank, displaced individuals. Neither shooting nor predation levels after the breeding season seemed to limit the breeding stock, and hence the next output of young, in the following year. Only if these mortality factors depressed the winter population below territory saturation could this happen. The two annual periods of displacement and increased mortality were thought to be the events which had previously been described as outbreaks of 'Grouse Disease'. Given that the displaced birds were more stressed and vulnerable than the territory holders, disease was probably often the final cause of their death.

This explained population changes within a single year. But what caused the variations in Grouse numbers over several years on any one moor and the differences in average populations between separate moors? The Grouse Team confirmed the earlier finding that the amount of good, young heather, as determined by the quality of moor management (especially patchwork rotational burning) was important to Grouse production. But their new finding was that stocking levels and breeding success varied markedly according to underlying geology. Areas of more base-rich rock were consistently more productive for Grouse than those for highly acidic rocks. The reasons are those discussed in Chapter 1 (pp. 24–26), which cause the base-rich soils to give heather of higher nutritional value to the Grouse (Moss, 1969). Stocks therefore depended on both quantity and nutritional quality of heather, and this explained much of the variation between moors in north-east Scotland.

Breeding performance clearly had a major influence on the number of birds available on 12 August on any moor and was much affected by nutritional status of the parent birds, which in turn depended on management practice, earlier weather and soil fertility (Watson and Moss, 1980). On certain Scottish moors infestation by sheep ticks carrying louping-ill disease caused serious chick mortality and contributed to population declines. Overall chick survival often showed little correlation with weather during the breeding season.

The Grouse Team found that population changes between years were caused less by differences in chick survival than by the adults' own behaviour. Aggressiveness changed according to population density. During population declines territory size was larger than during increases, suggesting that at peak densities the territory holders became more aggressive, and repelled their neighbours farther. More of the subordinate Grouse were thereby displaced from the breeding population than before, so that overall numbers declined. It was thought that decrease in food availability and/or quality adversely affected the female Grouse's nutrition and the next clutch produced chicks with lowered viability, and this was confirmed experimentally by hatching eggs in incubators. Chicks from lower-quality eggs were not necessarily more aggressive, though, and differences in aggression were evidently inherited. While decline involved a high rate of emigration from the population, increased aggression began only after the population had started to decline (Moss and Watson, 1980).

Cycling in numbers may thus stem from adverse extraneous factors, such as disease and food shortage, but also involves changes in inherent social behaviour not fully understood but presumed to have survival value for some individuals.

The statistic of greatest concern to shooters is, obviously, the size of the 'harvestable' stock at the end of each summer, which portends the size of the coming bag. A separate study of Red Grouse by the Game Conservancy in northern England focused first on historical data from game bags. These showed an average cycle in Grouse numbers of 4.8 years, compared with 6–7 years in north-east Scotland (Potts, Tapper and Hudson, 1984). In northern England infestation of adult Grouse by *Trichostrongylus* depressed breeding performance sufficiently to produce decline and to account for the observed cycles. Hudson (1986*b*) believes that such infestation can explain the increased aggression and dispersal, and the reduced breeding density, in the decline phases of cycles. Louping-ill was also found to be responsible for local or periodic Grouse declines.

The study of game bags indicated a general long-term decline in northern England Grouse populations since around 1900 (Fig. 32). There was wide geographical variation in population trends within this region, but largely unrelated to differences in geology or climate, and neither parasitism nor disease seemed to account for long-term declines. The factor most closely matching average bag size was the number of gamekeepers within 10 km of the moor. In the Peak District, a decline in average Grouse bags to just over one half between 1935 and 1980 was correlated with a comparable reduction in gamekeepers, and also with the tre-

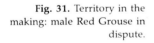

Fig. 31. Territory in the making: male Red Grouse in dispute.

bling of sheep numbers and consequent 30% loss of heather-dominated vegetation. The effect of keepers is thought to be through their control of predators such as Foxes and Crows, while loss of heather reduces the feeding value of the moor (Hudson, 1986*a*).

Game Conservancy study of bag records for grouse moors in Scotland also shows some evidence of long-term decline. In general, peak bags after 1939–45 never returned to their higher pre-war level, and in the South-West and the Central Highlands many moors showed severe decline from 1976 to 1983 (Barnes, 1987). A strong recovery in Grouse numbers on many Highland moors in 1987 led Watson and Moss (1987) to interpret this decline as an unusually prolonged and low trough of a short-term cycle. Hudson and Renton (1988) believe that, in parts of Scotland, winter predation limits the size of subsequent breeding population and may be a major factor in some long-term declines.

Even before 1940, the best Scottish moors never reached the extreme Grouse densities achieved in northern England, which were highest of all on moorland over acidic Millstone Grit. Most of the moors where high bags (80+ birds/km^2) have been maintained since 1976 are in northern England, and concentrated on the Yorkshire Dales. The most obvious difference between the two regions is in temperature: conditions will, on average, be slightly colder (by 0.5–1.0 °C) in comparable situations on the Scottish moors. Any effect of such climatic differences may have been reinforced by the decline in mean temperatures since 1950–60 and in particular the tendency to colder springs. Lamb (1982) says that in England the growing seasons (duration of temperature above 6 °C) have been on average about 9–10 days shorter annually since the mid–1950s than in previous decades. While he was referring to low altitudes and farm crops, a parallel effect could be expected for moorland vegetation at higher elevations. If temperature has a significant effect on the ecology of Red Grouse, then conditions have become less favourable in northern England, but still more so in Scotland. Another remarkable feature is that many of the best grouse moors in northern England were subject to enormous outfall of atmospheric pollution at the height of heavy industrial operations in nearby cities.

In at least some areas, large declines in Grouse numbers since 1900 have coincided with appreciable loss of heather-dominated vegetation as a result of increased sheep stocking and indiscriminate burning. The moors at Roan Fell, where the record Scottish bag was made in 1911, had become markedly less heathery in the 1960s compared with the 1920s (Fig. 27). Heather ground of Skiddaw Forest and Fauld's Brow in northern Lakeland has deteriorated considerably since 1944: sheep have increased and Grouse numbers are now at a low ebb. Many former Welsh grouse moors have become much grassier through the ascendancy of sheep management and probably fewer than 1000 Grouse now survive in Wales (R. Lovegrove, personal communication). A still more obvious cause of catastrophic decline is afforestation and large areas of grouse moor have been replaced by conifer forest in Wales, the Cheviots, the Southern Uplands and around the southern and eastern fringes of the Central Highlands. Moor burning has also ceased on many areas adjoining the new forests.

Another effect of afforestation may well be increased predation on the nests and chicks of species such as Grouse, extending outwards in a zone from the forest edge. Not only are predators such as Foxes and Crows

Fig. 32. Red Grouse population changes in two different districts. The graphs are based on the average number of grouse shot on a selection of moors in each district (a) and (b). The broken line shows the average number shot in each year, and reveals the tendency to short-term cycling. The solid line shows the average number shot for each five year period and, by 'smoothing' the fluctuations, reveals the underlying long-term trends. Reproduced with permission from Hudson (1986a).

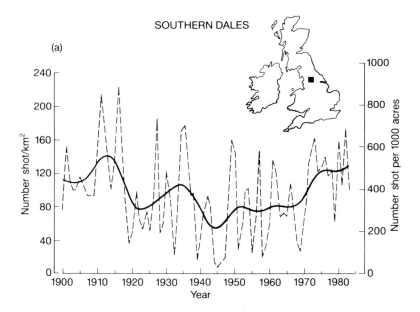

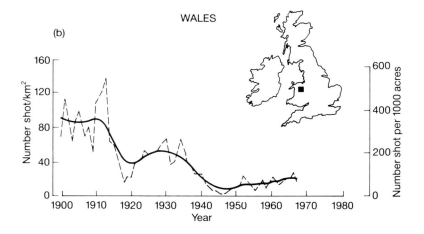

extremely difficult to control within the young thicket forests, where they find an ideal breeding refuge, but the change in land use may result in disappearance of keepers from the area. Non-breeding parties of Crows and even foraging gulls which have no direct connection with the forest may then become added (R. Parr, personal communication). The forest dwelling predators have to feed mostly outside, because little food is available within the thicket plantations. Marquiss and Newton (1982) noted that some Goshawks nesting in hill conifer plantations feed mainly on Grouse on adjoining moorland. Many north-eastern Scottish moorlands where Grouse declines have occurred now have extensive plantations in their vicinity. Conversely, the Pennines, where the best-stocked moors are now located, are the one large upland district where afforestation has made little recent penetration, because of the extent of common land.

The Red Grouse story is thus complex, and pieced together over decades by the painstaking efforts of dedicated researchers. Yet, even now, not all the mysteries are solved. The different dynamics of populations in separate areas seems to reflect differences in environmental influences. Rapid and regular cycling in numbers appears to be characteristic of areas optimal for the Grouse, where they increase rapidly and within a few years pass a critical threshold of population density at which a particularly strong peck order interaction occurs, so that a massive exodus and ensuing high mortality follow, and the population crashes. This behaviour has presumably evolved as an innate response to over-crowding, through giving the more dominant individuals, which stay, better changes of survival than if high density persisted or increased still further (Moss and Watson, 1985). It may be that both amplitude and frequency of the population swings induced by a normal control mechanism become exaggerated under the artificial circumstances of Grouse management, so that pronounced and regular cycling is not a really natural condition. Adverse factors, such as disease, may then come into play or even precipitate the heavy mortality of the crash, so that disease then appears as the obvious cause of death. In unfavourable environments, where inclement weather, poor food supply or heavy predation curtail the rate of increase, it may take Grouse longer to reach the critical density at which their own response produces a decline, and in some areas they may seldom if ever reach this threshold. This could account for the varying length and regularity of population cycles, or the local absence of cycling altogether. Variations in cycle length on one moor could be explained by chance annual differences in environmental factors, such as weather, which affect the time taken to reach threshold population density.

Long-term decline seems to differ from short-term cycling mainly in reflecting environmental deterioration with time. The enormously high Grouse densities reached on many moors before 1940 appear to have resulted from practices which maximised conditions favourable to the bird. The 'saturation' of many parts of the country with gamekeepers, and their quite ruthless destruction of any suspected predator, must have kept predation at very low levels over large stretches of upland. The large number of keepers also ensured careful rotational burning and the optimum management of many moors. And sheep were mostly at only low to moderate numbers. The position has been reversed since 1940, with decrease in keepers, and increase in both predators and sheep. Habitat deterioration through increased grazing pressure is bound to cause Grouse decline beyond a certain point, whilst the losses of both habitat and Grouse from afforestation are beyond dispute. Climate was perhaps also more favourable before 1950. The particular pattern and degree of environmental deterioration appears to vary widely geographically.

There is thus no simple, all-embracing explanation of the population changes in Red Grouse. The only general conclusion to be drawn here is that truth has many sides.

Black game

BLACK GROUSE overlap with Red Grouse, but belong more to the moorland edge, or to open hill woodland. They were once particularly birds of the 'white' ground where extensive grasslands prevail, especially of flying bent, or mixtures of grass and heather, with abundant rushes in the wetter places and bracken in the drier. In some areas they frequent the

Fig. 33. Blackcock at the lek: two males in mock aggressive but spectacular display. Robert T. Smith.

lowland peat mosses and their edges, and are often seen on the enclosed marginal land. Chapman (1907) found them nowhere more abundant than in the broad valleys of the Cheviots, where there was much wet, rough grassland on the lower moors, and many winding little branch glens with streams fringed by birch, alder, hazel and willow. In its wider Eurasian range, the Black Grouse is mainly a bird of the more open parts of the Boreal forest, of conifers and birches. They often perch in trees, while feeding and for roosting.

Blackgame are largely vegetarian, feeding on heather and other dwarf shrubs; buds, catkins and foliage of trees and shrubs; berries and fruits (including haws and rose hips); seeds, herbs and flowers. Insects are a minor item of adult diet, but are important for the chicks. In the uplands abutting arable land, blackgame often descend to the fields in autumn, to glean on cereal stubbles or even on potatoes and swedes. They are largely sedentary but descend to lower levels during deep snow. This species seems not to tunnel under snow, as do Red Grouse, but will shelter below for days if snowed over. Flocking in winter is usual, and though large parties are seldom seen nowadays, they once commonly numbered 50, and even up to 100 birds on the Borders.

The Black Grouse is best known for its lek behaviour, a spectacular performance much depicted in bird films, and so nowadays familiar to many. It has been described many times (e.g. Philipson, 1954; Bannerman, 1953–63) and Cramp and Simmons (1980) have given a detailed summary. Essentially it involves elaborate, ritualised and aggressive displays between males on a communal parade ground (the lek), and determines a peck-order in relation to mating. The preferred Blackcock lek sites are open areas of grassland on better soils where sheep graze selectively and keep the turf extremely close-cropped.

The birds assemble here at dawn, and sometimes at dusk or even in the middle of the day. Up to 40 will collect from a wider area, though the number is often much smaller, especially if only males are present. A variable number of cocks will take the stage at any one time, but each usually picks a particular sparring partner. Strutting back and forth with tail fanned over the back, white tail coverts fluffed, wings drooping and neck feathers ruffed out, they may sidle around each other for a time. Then two will lunge head-on, crowing hoarsely and sometimes leaping in the air, clashing with a flurry of wings, as if attacking in deadly earnest. Often these are only mock battles in which no blows are struck and the combatants withdraw with dignity and hardly a feather shed. In between these jousts they croon to themselves a far-carrying bubbling call which comes eerily over the moor to puzzle the uninitiated.

The females may sit, apparently uninterested, around the edges of the lek, a passive audience to this display of male aggression. Now and then a hen will enter the lek, ignoring several males, to seek out and mate with a particular cock. For the lek area represents a system of small territories for the males, which thereby establish dominance relationships, usually with the top birds in the centre of the arena. The energetic displays arouse the females, which tend to select the dominant, central males for mating, and then later disperse to nest solitarily. Yet lekking often takes place with only males present, including early in the breeding season (March), when the Greyhens are sitting (May), after hatching (June) and even in autumn (October–November). It then seems to serve the purpose of maintaining the social hierarchy amongst the Blackcocks.

The Black Grouse is regarded as a good example of a bird in which a polygamous mating system, associated with lek behaviour, has involved sexual selection, with showy and conspicuous plumage evolving in the males, and contrasting with drab and cryptic colouring in the females, which carry the burden of rearing the young. Leks are often permanent locations and some have been known for 50–60 years, though others have become deserted. There may be several within a few square kilometres, but number and distribution depend on the size of a blackgame population.

Though absent from Ireland, Black Grouse were at one time widespread in Britain from Cornwall and Devon northwards, and were especially numerous in parts of the Pennines, Cheviots and Southern Uplands, up to the 1920s and 1930s. Various observers noted that during the late 1930s there were signs of population decline, and by 1945 this had developed into a full-scale crash (e.g. Philipson, 1954). Abel Chapman (1907) surmised that earlier periodic declines on the Borders were the result of disease, but this has never been established by proper study. There were no obvious environmental changes to account for the decline, yet it seems to be generally true that the bird has never recovered to more than a fraction of its former population in its traditional habitats and areas south of the Highlands. In many districts young conifer plantations have rescued the Black Grouse, and become its main habitat. In the Cheviots and Southern Uplands it has often flourished in the recently afforested areas, up to the thicket stage, which is unsuitable for nesting. This gives a hint that the cumulative effects of sheep-walk management may not have suited it. Some forests have suffered damage, through the birds' fondness for eating the buds and shoots of young conifers, especially during hard weather, and numbers have been controlled. Leks here are either beyond the forest edge or on unplanted areas within. Where a forest is all at the 15+ year stage, blackgame usually decline again, but they will colonise the clear-fells later.

Black Grouse are on the whole retreating, having disappeared from the south-west England and declined severely in Wales, the Peak District and

Lakeland (Lack, 1986). It was once a still more widespread bird, occurring in lowland counties such as Lincolnshire, Norfolk and Hampshire up to the early nineteenth century. Introductions boosted numbers and distribution in some areas, but sometimes failed. This is a bird whose ways appear not yet to be fully understood.

Waders

Although far less numerous than the Red Grouse, the GOLDEN PLOVER has equal claims to be considered our most characteristic upland bird. Perhaps even more than the other, it is the spirit of the moorlands and the frequent companion of the hill wanderer during spring and early summer. The two have similar distributions but the Plover occurs over a wider range of habitats. Golden Plovers nest widely on the acidic grasslands but the best Golden Plover breeding areas are mostly on heathery grouse moors, both on the drier kinds and on blanket bogs where heather shares dominance with cotton-grass and bogmosses. It is also an important bird of the great flat 'flows' so extensive in the far north of Scotland and western Ireland (Chapter 5) and quite typical of the high montane plateaux where Dotterel are sometimes companions. Although the highest breeding density known recently is on limestone grassland, this is an extremely localised and rather atypical habitat.

Fig. 34. Golden Plover at the nest on grouse moors in North Wales. Many nests in this habitat are placed, as here, in recently burned heather. Dennis Green.

Ecologically and genetically, this is a most interesting species, by reason of its divergence from the ancestral stock of Golden Plover, which breeds through most of northern Europe and north-west Siberia (Fig. 35). There are related taxonomic problems and confusions. The European Golden Plover *Pluvialis apricaria* is regarded as specifically separate from the American–Asiatic Lesser Golden Plover *P. dominica*. Voous (1960) believes that the European form has evolved from the other through isolation, but the two have since developed an area of overlap, with reported hybridisation, on the Yenisei tundras in Siberia. The European species was once separated into two subspecies, on the basis of plumage differences, the Southern *apricaria* and the Northern *altifrons*. The extreme plumage forms are certainly very different: the southerner with little black on the face and neck and a rather mottled black breast, and the northerner a much more handsome bird, with intensely black face, neck, breast and white marginal band. There is every gradation between the extremes, and a rather complex cline in the frequency of these graded forms from south to north.

At the southern breeding limits, in Britain, Ireland, Denmark and north Germany, the paler forms predominate, while with increasing distance north through Scandinavia the positions become reversed, with a clear prevalence of really dark Plovers in the Arctic. Within Britain, such dark birds (seldom quite as black as the typical Arctic forms) can appear in any breeding area, and they are nowhere more than a small proportion of a population. The problem is compounded, because all these forms become indistinguishable in their winter plumage, so that the relationship then between immigrant or passage Northern European and the British–Irish populations (wintering in the lowlands) remains quite obscure. The subspecies distinctions have thus been abandoned, and within the single

Fig. 35. World breeding distribution of Golden Plover (re-drawn from Voous, 1960).

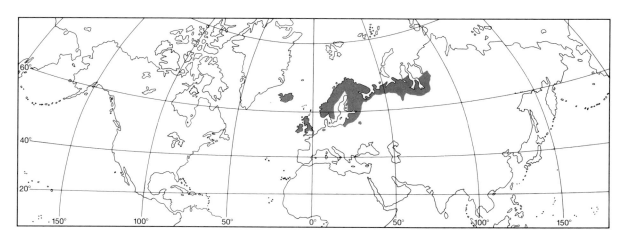

undifferentiated species of Golden Plover one should speak only of southern and northern populations (Cramp and Simmons, 1985).

Yet another intriguing issue is that the males are markedly darker and brighter than the females. In his study of a colour-ringed Golden Plover population in Kincardineshire, Parr (1980) graded birds on a scale of increasing darkness from 0 to 7. He found that all males belonged to grades 5 to 7, and all females to grades 1–4, and that there was a clear tendency for the darkness of the male to be correlated with that of the female paired with him. First year birds were sexually mature, had as much black on the front as the older ones, and showed no increase in amount of black with age.

The British–Irish breeding populations of Golden Plover appear to be genetically distinct from the Scandinavian, since there is no evidence that any of our birds move to more northerly lands in spring, or that Scandinavian birds stay here to nest. Out of 23 ringing recoveries of British-bred birds, only two were abroad: France and Portugal. The rest were recovered at an average of 115 km from their birthplace (Cramp and Simmons, 1983). Britain and Ireland thus have self-contained breeding populations adapted to our insular conditions. Although small numbers of Golden Plover in Northern Europe nest in open forest heaths and bogs, it there belongs essentially to ground above and beyond the forest zone. It is a wader of the wet or dry tundras and the barren fell-fields well into the Arctic, feeding and nesting within this treeless terrain. A few Golden Plover formerly nested in open forest habitats of the Cairngorms, but in Britain it is also largely a bird of treeless upland. In many breeding areas, Golden Plover feed especially on the better pastures of the enclosed fields just beyond the edge of the moor, or on the improved ground still remaining unenclosed.

The good pastures are often conspicuously the haunt of Plover through the nesting season. Flocks usually appear there when they first return in February or March, and there may be much activity, including courtship, pairing and display. The pairs then take to the open moor, there to stake out nesting territories, when conditions allow. They may come and go between the moor and the fields, but the beautiful song display by the males occurs mostly over the nesting grounds. Patrolling usually high over the moor, and sometimes approaching rivals near his invisible boundary, the male keeps time in slow wing beat to his far-carrying cry, 'tirr-pee-oo'. It is one of the most evocative sounds of the uplands. At the instant the song stops, the wingbeats quicken to their normal speed, and the bird assumes its typically rapid flight. Quite often it will plane to the ground, there presently to greet the intruder with the familiar single and plaintive alarm whistle.

The female will usually be in evidence also, before the eggs are laid or the clutch completed, and the pair hold the territory, alarming at humans and defending against trespassers of their own kind. This is the best time – in early or mid-April – to form an idea of the population density and size by walking over the moor. Once incubation begins, the pattern changes. The sexes incubate in turn and the off-duty bird typically retires to the favourite feeding place, to join company with others. The male usually sits during the day (Parr, 1980). The moor can often seem silent and deserted, with hardly a Plover on the place. Many sitters will stay firm and flush only when a person chances to come within a few metres, and even then they may keep silent or call little and from a distance. Yet in some areas, or under particular conditions, most incubating Plover will leave their eggs while a human is still at some distance (up to 500 m), usually by flying away low straight from the nest.

Some Plover appear to fly up to several kilometres from their nests to the favourite feeding grounds, but those near the edge of the moor may have to travel only short distances. On extensive moorlands, pairs deep in the interior may resort to more fertile grasslands, marshes and flushes on the moor, at least at times. The fields beyond the edge of the moor often show a remarkable constancy in use, the same selection being preferred over many years, whilst others of essentially similar appearance and situation are largely ignored. They are usually grazed by sheep or other stock and have a short mixed grass sward over a relatively fertile brown loam soil containing abundant earthworms, which are an important food item for the Plover. This is often Lapwing habitat, too, and there may be nesting pairs quite close to where the Plover feed.

Once the eggs have hatched, behaviour changes again. Both parents are usually in noisy attendance, often coming to meet intruders, and accompanying them until they have reached a safe distance from the brood. Adjacent pairs sometimes join up and form a group of scolding birds, and it is easy then to over-estimate the actual number of breeders. Both parents, as well as the chicks, usually now have to depend on the nesting ground for food. Sometimes the whole family will move quite quickly to better feeding places on the moor, including grassy or marshy ground, where Plover do not nest, but until the young have fledged they do not usually appear on enclosed pastures, except when these are very close. The feeding requirements of the chicks, and possibly also of the adults, probably dictate the clear preference for short vegetation in which to nest. There is, accordingly, an important connection between suitability of the ground for nesting, and the way it is managed.

Long heather and tall tussocky cotton-grass or flying bent are mostly avoided by nesting Golden Plover. The invertebrate prey is more hidden and is it more difficult, especially for the chicks, to move through such long vegetation in search of food. On grouse moors the Golden Plover thus benefit from regular rotational burning. They nest mostly on ground burned within the past 8 years, and on moors with large blocks of old heather their numbers may be held below the level which the ground could carry if there was a small-scale pattern of burning. Release of grass-land from fire and grazing can have the effect of displacing Plover by allowing the sward to become dense and tall. On naturally treeless nesting habitats, such as the montane heaths, grasslands and fell-fields, and the *Sphagnum*-dominated blanket bogs, vegetation remains short and suitable for the bird under the prevailing climate.

Though there can be occasional clusters denser than the average, Golden Plover tend to space their nests fairly evenly over the moor when all the ground is suitable. On many of the better breeding areas, pairs and their nests are spread out at about 400 m between nearest neighbours, which gives a mean density of around 5–8 pairs/km^2. This level is seldom exceeded on acidic moorland, but densities of 10 pairs/km^2 are reported from heather ground with a small-scale pattern of rotational burning in the Moorfoot Hills (L.H. Campbell) and from earlier records may have been surpassed in the Peak District before 1940–50. In some areas breeding densities of 2–3 pairs/km^2 are more usual, and on many barren moorlands in the western Highlands and west Ireland, 1 pair/km^2 or even less is quite common. The highest density in Britain recorded in recent times is 16.4 pairs/km^2 on close-grazed, hummocky limestone turf with mat-grass patches in the Pennines (Ratcliffe, 1976).

In some areas it appears that the Golden Plover fields beyond the edge of the moor have parties of non-breeders through the nesting season, as well as the off-duty nesters. This suggests that numbers of breeding pairs are limited by their territoriality rather than the size of the over-wintering population. Parr (1979) made the interesting observation that on Kerloch Moor, Kincardineshire, there was frequent sequential nesting of Golden Plover, with new pairs moving in to breed in territories of others which had failed or hatched and moved elsewhere. This showed that some of the apparently non-breeding pairs in the fields were simply waiting their turn to nest, and that breeding population can be larger than the number of nests at any one time would suggest. We do not yet know how frequent or widespread this curious behaviour may be.

While Golden Plover are thus far less numerous than Red Grouse their breeding behaviour shows certain similarities. Interestingly, their densities also tend to run in parallel: moors good for the one species are usually good for the other, and vice versa. In some parts of Scotland, numbers of Moun-tain Hares also tend to match those of the two birds. Yet, by contrast, the Plover are not prone to the wide numerical fluctuations so characteristic of

Grouse. Some observers report annual variations in breeding strength on certain haunts, but in the areas I have examined, the relative constancy in breeding density between one year and the next has been quite striking. There is, nevertheless, convincing evidence for a long-term decline in Golden Plover populations in some districts, such as Wales, the Peak District, the Cheviots, parts of the Southern Uplands, parts of the Highlands and Orkney (Ratcliffe, 1976). In Ireland the species has both declined and retreated northwards, having disappeared from the nineteenth century breeding haunts in Wicklow, Tipperary, Cork and Kerry (Sharrock, 1976).

Afforestation is the most obvious and serious of the causes of decline (pp. 214–218). This bird is one of the first to disappear when the nesting habitat is ploughed and planted with conifers. There is little chance of more than an occasional pair of Plovers returning to clear-felled areas of forest before the next generation of trees grow up. Even areas of moorland left unplanted within or above the forest usually lack the species unless they cover several square kilometres and remain grazed: the vegetation simply grows too rank. The open moorland beyond the forest may also become inhospitable through lack of burning, since this is discouraged because of fire risk. This may be one reason why Golden Plover and other waders tend to avoid the forest edge over a perimeter up to 800 m wide (Stroud et al., 1987). Another contributory factor may be the potential for increased predation resulting from increased refuge for Foxes and Crows in the forest, and reduction in keepering. Although there was no study of a 'control' area, Ray Parr found that afforestation of half of Kerloch Moor resulted in parallel decline of the Golden Plover breeding population, and that breeding success in the remaining segment declined through increased nest predation, by Crows and Gulls.

This accords with the observation that most good Golden Plover populations are on heavily keepered grouse moors, and that declines in some areas have followed the abandonment of grouse management. On some moors nest losses to predators are usually heavy, and although most will repeat at least twice if necessary, the eventual breeding success is low. Some individuals or pairs of predators such as Crows seem to become skilled at finding nests and can account for the majority of clutches over a wide area. Golden Plover do not show aggressive defence of their eggs or young in the same way as Lapwings and Curlews, and may be at a disadvantage where nest predators are numerous.

Other southern populations of Golden Plover in continental Europe south of the Baltic have declined markedly: the species no longer breeds in Belgium, the Netherlands or Poland, but a handful may hang on in West Germany and Denmark. Habitat loss is believed mainly responsible, and even in southern Sweden the bird has declined seriously through draining and afforestation of its peatland and heathland habitats.

The wetter grouse moors, with well-developed blanket bog, are the most typical upland haunt of the DUNLIN. This bird is especially associated with bog pools or peaty moorland tarns in its feeding, but also resorts to stonier and richer lake and river margins. Dunlin seldom nest on dry heather moor or upland grassland, and so have a much more restricted distribution than Golden Plover. There is a small outlying group on Dartmoor and the bird was once widely distributed through the peatier uplands of Wales. The Pennines are one of the main haunts, with a fairly

Fig. 36. Breeding distribution of Dunlin in Britain and Ireland during 1968–72 (from Sharrock, 1976).

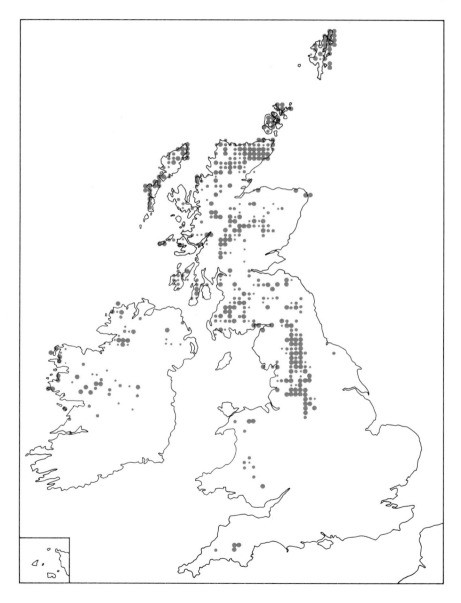

continuous distribution from the High Peak to the Tyne Gap. Dunlin were also once widespread along the Cheviots and on the more gently contoured areas of the Southern Uplands.

In the Highlands, many pairs nest on the high montane plateaux of the central massifs, but the pool and hummock bogs of the Flow Country are the great upland stronghold of the Dunlin in Britain (Chapter 5). The species is still widespread in Orkney and Shetland, but evidently less numerous than 50 years ago. It is also widespread on the peaty moorlands of the Outer Isles, especially Lewis, but the population density on the marshy parts of the coastal machair here far eclipses those in any other British habitat.

In Ireland it occurs rather sparsely in the west and north-west, mainly on blanket bogs and wet moorland, but also in richer coastal and lowland marshes. Small numbers occurred until recently on the raised bogs and other wet ground habitats in the central Irish plain. Several races of this circumpolar wader are recognised, the British breeding population belonging to the southernmost form, but lacking the degree of plumage differentiation shown by the southern population of Golden Plover.

Hill Dunlin are little studied, for their numbers are usually sparse and difficult to count with confidence, and their nests are elusive. They usually return directly to their often wild and remote moorland breeding grounds in spring. Pairs or threes or fours will suddenly appear weaving their way at high speed through peat haggs with a thin, sizzling call. Alternatively flashing pale underparts and darker back as they zig-zag close to the ground, they are quickly gone again. Or there will be single birds or pairs sitting attentively on tussocks and giving a short churring trill. The nest will later be in a tussock, usually quite well hidden, and the neatest little cup, lined with cotton-grass leaves and containing four eggs. They vary between a beautiful pale green and pale olive-brown, variously blotched and speckled with reddish brown, and the markings often have a spiral twist.

All waders' eggs are cryptically coloured, but some species suffer considerable egg predation nevertheless, and even when the nests are well concealed, some predators can still find them. A balance develops in the contest, but if the predators begin to get the upper hand, breeding output of the predated species may be so reduced that their decline is inevitable. While predators cannot afford to deplete their overall food supply, wide-spectrum feeders such as Crows may be able to prey selectively on a minor food item, such as eggs of a particular bird, to the extent of causing a serious impact on that species, but without themselves being adversely affected.

The moorland Dunlin nearly always have the company of Golden Plover, and they have developed this companionship to an intriguing degree. A watchful Plover on a tussock in its nesting area will often have a Dunlin standing beside it. When the Plover flies, the Dunlin will at once rise and follow close behind, to pitch again within a few metres of its neighbour. Often, when the Plover calls, the Dunlin trills. Sometimes up to half a dozen birds will attend upon one or more Golden Plover in this way. This behaviour long ago earned the Dunlin the name of 'Plover's page' (Figure 37). Occasionally, they will form the same attachment with Greenshank and Dotterel. It is supposed that one or both species gain some advantage from this association. Thompson and Thompson (1985) found that Golden Plover and Dunlin in mixed association flew at appreciably greater distances on the approach of intruders, compared with either species when only in the company of its own kind. They deduced that this greater responsiveness to the approach of potential predators may reduce the risk of predation, at least for the Dunlin.

Dunlin are not, on the whole, shy birds, and when the eggs have hatched they often become quite confiding, standing on tussocks close to the intruder and flitting or running quickly along to keep just ahead, purring in an anxious way. It is a form of distraction behaviour and may give way to injury-feigning to draw pursuit. The chicks meantime crouch hidden in the vegetation until the parents signal that the coast is clear. They are beautiful little mottled balls of fluff and adroit at picking up moorland insects for themselves soon after hatching. They and the adults depend especially on

Fig. 37. The Plover's Page: Golden Plover and Dunlin.

the diptera, notably the midge family, which thrive in wet ground and particularly bog pools, but also craneflies. The Dunlin is a small wader and with its short bill needs to find food on or just beneath the ground surface, or in shallow water. The quality of moorland habitats for it seems to depend especially on the abundance of pools and wet *Sphagnum*-filled hollows. The best areas in the Flow Country for the bird are where there is a high density of rather small and shallow pools or *Sphagnum* hollows. On the Alaskan tundras, Holmes (1970) found that Dunlin breeding density was correlated with abundance of Chironomidae (midge family) which in turn was related partly to frequency of pools.

The Dunlin is a good example of a wader whose breeding density and habits vary according to food supply. In Britain and Ireland there is less detailed information about food, but a very clear connection is known between the extraordinarily high density of nesting Dunlin on the Hebridean machair marshes and lochs, and the highly calcareous nature of the shell sand substrata and associated standing water. The best moorland habitats cannot compare with this and are mostly on acidic peatlands. Over much of its upland range the bird is at only low to moderate densities (1–5 pairs/km^2), but on good pool areas numbers sometimes reach 10–15 pairs/km^2 and the bird becomes semi-colonial, with spacing between pairs of 200 m or much less. On some of the machair marshes, normal territorialism seems to break down and the Dunlin becomes virtually colonial. One South Uist machair had the incredible density of 307 pairs/km^2 in 1984, but the average over this and 7 other machairs in 1984 and 1985 was 27 pairs/km^2: the total Hebridean machair population is estimated at *c.* 3800 pairs on average (35% of the British–Irish population) (M. Pienkowski and D. Stroud). Holmes found that in Alaska, densities varied from 75 pairs/km^2 on subarctic tundra to 15 pairs/km^2 on high arctic tundra 10° farther north.

Some breeding populations of Dunlin appear prone to annual fluctuations. In the northern Pennines, they seem to be influenced by weather preceding the laying period. Prolonged drought, especially with prevailing cold east winds, during March–April leads to drying out of many of the feeding pools on moorland. A cold, backward April and May will also retard the cycle of insect abundance and, even though the ground remains wet, may have the effect of discouraging many Dunlin from nesting. In some moorland areas there appears to be evidence of long-term decline even though the habitat has changed little in appearance. Nesting has long ceased on the lowland peat mosses bordering the Solway, where small numbers bred in the nineteenth century, and the bird has almost gone from the salt marshes of this and other estuaries where it once nested at higher density than on most uplands. An obvious and important cause of upland decline is afforestation, which has destroyed many breeding haunts (see pp. 214–18). The Dunlin has also declined on the European continent south of the Baltic, partly through obvious habitat loss.

The Curlew is a widespread and constant bird of the grouse moors, though perhaps seldom as numerous there as on some grasslands. Lapwings are patchy and mainly in the lower and grassier areas, while Snipe and Redshank are widely distributed. The upland stream and lake birds of the sheep-walks are usually well represented, and moorland tarns are often the location of Black-headed Gull colonies, varying from a few to several thousand pairs. All five of our inland nesting species of gull nest on moorland flows (see Chapter 5).

Song birds

The hill passerines are an important bird group on the grouse moors. My impression is that Ring Ouzels reach higher numbers on the heather moorlands than in any other uplands: they have a greater abundance of berry-bearing plants than on the grasslands, and this must be an important late summer source of food. In the eastern Southern Uplands, Poxton (1987) found that Ring Ouzels were strongly associated with heather slopes.

The STONECHAT also finds especially congenial habitat on heather ground and occurs more commonly here than on most sheep-walks. A majority of upland Stonechats move to the coast and southwards during winter, for conditions are much milder here and there is far less of the frost and snow which can produce such heavy mortality. Some pairs nevertheless remain on the hills the year round, and Bolam (1913) noted them in bleak, elevated and mainly stony winter haunts in the Welsh mountains. These birds are greatly at risk in severe winters, and even those which have moved to more favourable locations may be decimated then. The bird tends to have irregular cycles of numbers in the uplands, building up and spreading, only to crash and almost disappear from many areas during the next severe winter. In some of the northern Lakeland fells, its place appears to be taken by the Whinchat for a few years after a crash, for this summer visitor is able to escape severe winters and is less prone to large swings in numbers.

Perhaps the Stonechat's optimum habitat is the maritime heaths of heather and gorse which occur especially in western Britain and Ireland. These become the main breeding refuges after heavy winter mortality and, with often three broods a years, numbers soon build up again to allow recolonisation of depleted areas inland. The Stonechat is noted for its fidelity to particular nesting sites during periods with mild winters, and even after a gap, pairs often reappear in previously favourite haunts. A species of conspicuous habits, this is one of the easier small upland birds to count and keep under annual surveillance.

The TWITE used to be regarded as a typical bird of the heather moorlands, but during the past 100 years it has declined mysteriously, sometimes to vanishing point, in many hill areas, in north Wales, Lakeland, the Cheviots, Southern Uplands, and some parts of the Highlands and Islands. Former lowland peat moss nesting haunts in Lancashire and on the Solway are now mostly abandoned. Nowadays, as a breeding bird, it is found especially on the western and northern seaboard of Scotland and Ireland. Inland in the Highlands it is irregularly scattered over the grouse

Fig. 38. Breeding distribution of Twites in Britain and Ireland during 1968–72 (from Sharrock, 1976).

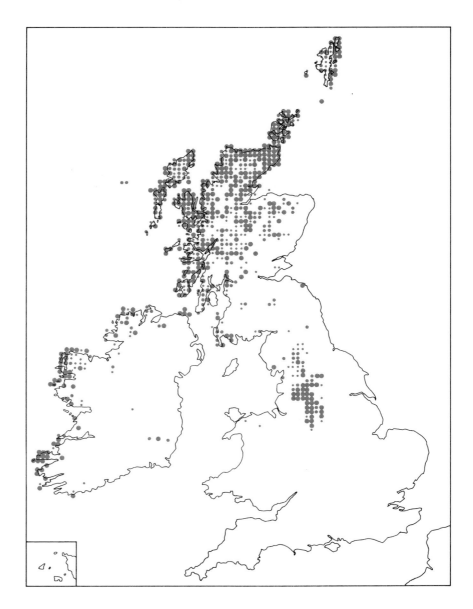

moors and deer forests, but in some areas is found nesting especially on vegetated cliffs.

There is, however, a remarkable stronghold in the central Pennines and Peak District, where the Twite is mainly on dry moorlands covered with heather, bilberry and bracken, with cotton-grass blanket bog on plateaux above. The nesting places are mostly in the denser vegetation of this range of habitats, but the birds commute to feed mainly on the upper farm pastures and meadows. The feeding grounds are usually about 1.5 km from the nesting areas, and the young appear to be fed entirely on regurgitated seeds (Orford, 1973). Some food is obtained from the moorland, especially from heather, by the hen during incubation. Burned *Molinia* grassland is feeding habitat between the birds' spring return and the onset of nesting. The Twite has a spread-out breeding season, with birds still returning to their nesting grounds during June, though many pairs have eggs in May (Orford, 1973).

Twites begin to flock and desert their upland nesting grounds in autumn, often moving across the hills as they head for wintering haunts mainly scattered around the coasts of Britain and Ireland (Lack, 1986). Only occasionally or in a few areas do Twites winter amongst the hills. The present distribution of the Twite, and its scarcity or absence from many apparently suitable moorland areas is a puzzle. The bird breeds in a wide range of habitats, yet is now concentrated into two main populations: the western coastal in Scotland and Ireland, and the Pennine moorland. Sharrock (1976), in an interesting summary of the history, distribution and ecology of the species in these islands, points out that this discontinuity is a smaller-scale counterpart to the large disjunction between the central Asian and Norwegian–British Twite populations. This appears to be a bird which has retreated – for reasons unknown – from a more continuous range in Britain and Ireland, and seems unsuccessful in its scattered attempts to recover lost ground, despite its good status in its local strongholds. Davies (1988) gives reasons for supposing that the Pennine Twites winter as a group on the salt marshes of eastern England between the Wash and the Thames, or on the coasts of the adjoining Low Countries, and this may help to maintain their distinctness as a breeding population.

The Twite is well named the Mountain Linnet, though here and there the Linnet breeds in close association with it on dry moorland. Orford (1973) found small numbers of Linnets nesting in heather among Twite colonies in the southwest Pennines up to 240–580 m, but noted that they had a distinctly poorer breeding success than the Twites, being evidently less able to withstand adverse weather conditions.

The NIGHTJAR is a mainly lowland bird which once occurred quite widely on moorlands over a large part of Britain and Ireland especially on bracken and heather ground, often in rocky and open places around the edges of woods or amongst open tree growth. Even blanket bog was amongst its habitats (Sharrock, 1976). The strange, reeling song of the male bird in the dusk was as familiar on some of the lower hillsides as on the southern heaths. Then, from the 1950s, the Nightjar declined rapidly as a hill bird, and is now found there in thinly scattered localities showing no clear geographical distribution pattern. It often favours clear-fells amongst upland conifer plantations, but is very patchy even here. The reasons for this retreat are unclear, and attempts to connect it with recent climatic changes are unconvincing. For migratory species such as this, the critical factors could be during the much larger part of the year when the bird is outside this country.

Predators

Whilst the grouse moors have most of the birds of the hill sheep-walks, the predators tend to be unwelcome, so that they are not notable areas for the Golden Eagle, Buzzard or Raven. Peregrines are more tolerated nowadays, but their relationship with Red Grouse is a sensitive issue. There are contrasting attitudes to the two most characteristic raptors of the grouse moors, the Merlin and Hen Harrier.

Heather moorland is the breeding habitat *par excellence* of the MERLIN. This, our smallest raptor, has been the subject of several recent studies, but still remains a somewhat puzzling bird, whose ways are not yet fully understood. It is a northern species, with a largely upland though submontane breeding distribution in Britain and Ireland, and is unknown as a nester elsewhere in Europe south of the Baltic. Exmoor was a well-known breeding area, but nesting on Dartmoor has an uncertain history, despite the large extent of suitable ground. Merlins were widespread amongst the Welsh hills and had a few coastal nesting haunts on rugged scarps and amongst marram on sand dunes. This was also a generally distributed species through all the hill country of northern England, Scotland and Ireland though evidently most numerous on the drier, heathery grouse moors of eastern districts. Nesting is unknown on the heaths of southern and eastern England, but formerly occurred on the lowland peat mosses of Cumbria and perhaps districts farther north.

Recent surveys have shown that Merlin breeding populations have declined appreciably in some districts since around 1950, or even since 1975 (Bibby and Nattrass, 1986): on Exmoor, much of Wales, the Peak District, Lakeland, the Cheviots, the Southern Uplands, Orkney and Shetland. Two other conclusions emerged from this study: that numbers were highest and holding up best on the main grouse moor areas of northern England and Scotland; and that declining numbers tended to be accompanied by poor breeding success. Ireland was excluded from the survey, but the Merlin evidently remains a sparse though widespread breeder, and still nests on the raised bogs of the central plain.

The obliterative effects of afforestation on moorland are clearly one major cause of Merlin decline in areas such as the Cheviots and south-west Scotland. While in other countries this bird can live in open forest and forest edges, it is not adapted to closed forest, and the growth of extensive thicket plantation removes its nesting habitat and food supply on a large scale. Some pairs may cling to traditional sites on rocky ground left unplanted within the forest, or along its margin, but they continue to take prey from the open moorland rather than the forest, and so depend largely on a sufficiency of unplanted moorland remaining beyond the trees (Newton, Meek and Little, 1978, 1984). It has also been suggested that Sparrowhawks which colonise the developing forests will displace Merlins in the vicinity (D.N. Weir, unpublished).

Organochlorine pesticide effects have been diagnosed as an important factor in some areas, with eggshell thinning at levels associated with significant reproductive failure (Newton, Robson and Yalden, 1981). There has recently been a rather puzzling discovery that the number of young raised by Merlins is inversely related to the levels of mercury in their eggs in mainland Britain, but not in Orkney and Shetland, where mercury levels in eggs were highest of all (Newton and Haas, 1988). This remains the British raptor most heavily contaminated by persistent toxic residues, and there may still be pesticide problems for some Merlins, which winter in lowland farm and coastal areas, since there is little exposure to contamination in most breeding areas. Yet Bibby and Nattrass (1986) concluded that there may also be inadequately understood factors operating against Merlins on some of the nesting grounds. They estimated that the British Merlin breeding population in 1983–84 was 550–650 pairs, but this was the first attempt at a national census and there are no earlier figures against which to measure the scale of decline. The true figure could well prove to be larger.

This is one of the most difficult upland species to census, requiring much hard effort for a reliable count even over 100 km². While pairs display and are conspicuous in their territories before laying time, they can be very elusive later. A pair may show interest in one part of the moor, but then move elsewhere to nest. When incubation has begun, many Merlins sit until a human intruder is very close, and at more than 25 m are almost impervious to hand-clapping or shouting. Incubation is shared between the sexes, with the male taking up to a third of the burden (Newton *et al.*, 1978), and the off-duty bird frequently leaves the nesting area for hours on end. It is quite easy to walk through ground with a nest and never realise Merlins are present unless one has the good fortune to find the plucking places, which are usually within 100–200 m of the nest. Even when Merlins breed successfully, they may escape notice if there is no prior knowledge of their presence. The parents are usually demonstrative when there are young in the nest, but sometimes not until the intruder is relatively close. And once the young are fledged the whole family can quickly move away from the vicinity of the nest.

The favourite nest site is on the ground, among fairly tall, dense heather. Steep, dry slopes with scattered rocks are favoured, and especially those in small, water-worn valleys. A few Merlins nest on the hummocks of flat flows where the outlook is far more restricted. Many pairs also resort to the unused tree nests of other birds – Crow, Magpie, Raven, Buzzard – either near the edge of upland woods, or in the last scattered trees and bushes on the hill slopes or gill sides. Some use crag ledges, either on open hillsides or in enclosed ravines, and the tops of huge detached blocks crowned with heather or bilberry. Nests of one pair in successive years are often close, but over a period some territories show use of a selection of alternative sites, spread over distances up to 2 km apart.

Merlins are territorial and often show the same kind of regularity in nest spacing as the Peregrine, with an average of 3–4 km between neighbouring pairs in good areas. Yet two adjacent nests are sometimes only 1 km apart, or even less, but with no others for many kilometres. In some areas there is an evident tendency to fluctuation in numbers between years, as well as a recent downward trend, e.g. in Northumberland (Newton *et al.*, 1986). This is reflected in the presence of non-breeding pairs, or only single birds, or by complete desertion, in some territories in some years. Breeding density has always varied widely, even before the recent decline. Bibby & Nattrass (1986) found that it was exceptionally as high as 5–10 pairs per 100 km², but only 40 out of 321 10 km squares with breeding Merlins had more than 2 pairs in one year. Many squares did not contain suitable habitat over the whole 100 km², though, and this reduced density measurements accordingly. The densest populations were on heather moors, where ground nesting predominates. Densities are generally low on the grassy sheep-walks, where tree-nesting is more frequent.

Annual variations in breeding strength would be partly explicable according to fluctuations in passerine populations on which the Merlins feed. Meadow Pipits are everywhere the favourite prey. In Northumberland, Newton *et al.* (1984) found them to represent 56% of breeding season prey by frequency, with Skylark next at 12%. In Wales, Bibby (1987) found Meadow Pipit prey at 61% fequency, and Chaffinch second at 8%. In Northumberland a further 48 and in Wales a further 52 bird species made up the remainder of the prey found, but no other species exceeded 5% frequency.

Hard winters and/or springs could have an effect on the overall populations and breeding performance of these small birds, to reduce available food supply for Merlins. Geographical differences in breeding density could similarly reflect differences in food supply between upland areas and habitats. Yet the evidence for this is far from clear-cut. Counts of small bird populations, especially Meadow Pipits, on grassy sheep-walks have proved not to differ significantly from those on heather moorland.

The most obvious difference between these two habitats, as various observers have commented, is that the ericaceous dwarf shrub heaths support considerable numbers of large, day-flying moths (notably the Emperor, Fox and Northern Eggar), whereas the grasslands hold very few. The larvae of these moths feed on dwarf shrubs, but the grass feeders are much smaller insects, e.g. Antler Moth. Direct observations and the fre-

Fig. 39. The Stone Falcon of the moors: Merlin on plucking block.

quency of wing remains at plucking places show that Merlins certainly take many of these moths, but whether in sufficient number to form a significant part of their diet is uncertain.

Merlins evidently decline when grouse moors become converted to grassy sheep-walks. The disappearance of good expanses of tall heather deprives the bird of this favourite nesting habitat, but on most sheep ground it can simply turn to disused Crow nests in bushes and trees. Bolam (1913) noted an increase in use of tree nest sites in Merioneth under heavier grazing by cattle, but he attributed the change in preference to the direct trampling disturbance for ground nesters. Newton *et al.* (1978) found that Northumbrian Merlin tree nests had a higher rate of fledging success (94%) than nests on the ground or on small crags (63%). Ground nests are vulnerable to mammal predators: Fox, Stoat, Weasel. Tree nests would only be safer in areas with little human persecution, though, because they are far easier to find than ground nests. Only one of the 100 Northumbrian ground nests was trampled, evidently by a sheep. If nest safety is a major factor, why has the Merlin not adapted more completely to nesting on ledges of bigger cliffs? Some Merlin crag sites cannot be reached without ropes, but these situations are only very locally used.

Gamekeeper persecution used to be a problem for Merlins on the Victorian grouse moors. Seebohm (1883) recounted that, on the Peak District moors, two nesting areas were reoccupied annually, although one or both birds of each pair were destroyed every year, and no young were reared over a 10 year period. In each area the same heathery slope acted like a magnet to new birds, though there was plenty of other apparently suitable ground in the vicinity. Enlightenment dawned slowly, but eventually most keepers came to accept that Merlins did little if any harm to grouse chicks, and the bird is left in peace on most moors nowadays. There are still some unfortunate exceptions, however.

During the earlier part of this century, a favourite egg-collectors' area for Merlins was the rather grimy section of middle Pennine moorland in the textile belt of Lancashire and Yorkshire. Here, nests were said often to be in bracken beds, sometimes on slopes with a mill below. A day's round of known haunts could easily give a tally of at least 3 nests (H. Buchanan). Merlins seem virtually to have disappeared from such areas, though it is unlikely that collecting itself was responsible. But what sustained this population before? A prolific food supply has to be assumed, and either there was an unusual abundance of Meadow Pipits or the Merlins had adapted to the town House Sparrows and Starlings. The latter is not too fanciful a thought when one considers that Merlins have lately adapted to nesting right inside various Canadian towns, feeding on these and other garden birds. More recent decline in the Peak District, where extensive moorlands remain apparently unchanged and suitable, has been attributed to pesticide contamination in winter haunts, for want of any other explanations (Newton *et al.*, 1981).

The most graceful of the upland birds of prey, the HEN HARRIER, is particularly, and often fatally, associated with the grouse moors. While this was evidently always mainly an upland and northern bird in Britain and Ireland, it seems in earlier times to have had a wide distribution which included lowland heaths and marshes. In mainland Europe and North America the Hen Harrier has a wider range of breeding habitats, including lowland fens which in Britain are more the habitat of Marsh and Montagu's Harriers. Nesting on the low-lying raised bogs was once widespread and Macpherson (1892) mentioned that three pairs nested in 1785 on Newtown Common, now part of the western outskirts of Carlisle. The older county avifaunas sketch the repetitive and melancholy tale, of a former widespread distribution on the moorlands, followed by a catastrophic decline under the impact of game preserver destruction, to virtual extinction by around 1900.

By this century, the Hen Harrier had retreated to the Outer Hebrides (especially the Uists and Benbecula), the Orkney Isles, possibly a few parts of mainland Scotland, and some of the remoter uplands of Ireland. Any resurgence during World War I was quickly mopped up afterwards and, apart from sporadic and usually unsuccessful nesting attempts elsewhere during the 1920s and 1930s, this remained its distribution. Orkney became its headquarters, and here protection effort was initiated by George Arthur, since the rarity of the bird was proving an irresistible magnet to egg collectors. World War II gave the Hen Harrier its opportunity. The absence of the keepers from the grouse moors, and the advancing plantations of the Forestry Commission, provided the much-needed 'break'. Donald Watson (1977) has traced the dramatic recovery and spread of the bird, which by the late 1960s had become restored as a widespread breeder in the uplands of both Britain and Ireland.

The two main nesting habitats make an interesting comparison. Older stands of heather on the grouse moors appear to be the favourite nest site, but in some areas the dense growths of common rush which dominate extensive areas of surface water seepage are used as well, and nests are quite often in mixed moorland vegetation. On heather moors, the nest sites tend to be moved around as the old heather is burned off year by year. In young conifer plantations, the nests may be in long heather or lush growth of grass such as *Molinia* with bog myrtle which have developed after burning and sheep grazing ceased. Gently sloping ground is favoured in either habitat, but some nests are placed in fairly steep slopes. Any particular forest breeding haunt is usually abandoned when the young trees close, but Donald Watson has known pairs in Galloway to continue breeding in thickets up to 5 m high, relying on the very limited space where a single tree has died or blown over. In these instances, survival of one or both of the parents was probably necessary to maintain this continuity of occupation. Where new forest is still being created, the birds will tend to move into it from older growth which has become unsuitable.

When the Hen Harrier gradually spread southwards from its northern refuges, the heather moorlands of east Sutherland, Moray, Perthshire and Stirlingshire were the first mainland areas to be recolonised, by about 1950. Thereafter, the spread was partly into other grouse moor areas, but especially south-westwards through the afforested areas of Argyll, including the Kintyre peninsula. By the late 1950s, the young forests of Kielder in the Cheviots and Galloway were colonised, but in the western Southern Uplands, some pairs bred on open moorland. The next expansion was in Wales, from 1962, where Hen Harriers spread over the heathery grouse moors of the north-east, but, strangely, have largely ignored the large areas of young forest. In Ireland a substantial recovery dated from about 1950, with the Harriers occupying both open moorland and young forest. The increase was mainly in the southern half of Ireland, and could have been a main source of the birds which reoccupied Wales.

During the 1960s and 1970s Hen Harriers nested in many moorland districts around the eastern and southern fringes of the Highlands, the Southern Uplands and northern England, and the species could again be described as a widespread breeder. Yet numbers seem now to have passed their peak and some areas report significant decline. Picozzi (1984) reported 85 nests on mainland Orkney in 1978; in 1981 he found 46 nests in the same study area, and decrease has continued, coincident with a great expansion of Orkney moorland reclamation. In the Kielder area, the small population disappeared when the young forests closed. Although a great deal of new forest has been established elsewhere in this district and across the Southern Uplands, it has either remained uncolonised by Hen Harriers, or only sparsely and locally utilised.

While breeding has been recorded in clear-felled areas of upland conifer forest (Petty and Anderson, 1986), it seems unlikely that this temporary renewal of open ground (it is usually quickly replanted) will prove nearly as attactive as the first ten years of the original plantations. Food again appears to be the problem. Donald Watson has found that forest nesting haunts are mainly where the harriers can hunt over adjoining unplanted ground with reasonable prey populations. Both territory occupation and breeding performance increase with the extent of low elevation moorland and marginal land available for hunting, for both open moorland and forest breeders. Large areas of young forest on moorland over infertile soils often fail to attract nesting harriers. This dependence on carrying capacity seems to account for the general absence of the species over large areas of the western Highlands and Islands, even where there are young forests.

The erratic distribution and numbers in the grouse moor country are most likely to be the result of persecution. This bird is disliked more than any other raptor by the grouse preservers and is widely, though quite illegally, destroyed on many moors. In some areas, harriers are killed as fast as they appear; in others they are picked off when the eggs have hatched and the nests are easier to find. Sometimes nests escape in remoter places or because the keeper onslaught is less determined or consistent. The failure of the bird to become more widely established in the Pennines can hardly be explained in any other way, and a flourishing population of 13 pairs in the Bowland Fells in 1983 was reduced to virtual extinction by 1986 through a coordinated keeper onslaught (P. Stott). This does not prevent new birds from continuing to try, and if persecution ceases then rapid colonisation and increase is possible.

There is no denying that Hen Harriers take considerable numbers of Red Grouse, both old and young, as prey during the breeding season. The food recorded at Galloway nests by Watson (1977) shows grouse to be the most important item by mass though it may be over-represented in frequency. In a study of breeding Hen Harriers on the grouse moors of north-east Scotland, Picozzi (1978) found that Red Grouse were the most important prey species by mass (55%) and lagomorphs second (34%), the young of both being the most frequently taken. The larger female Harriers took an appreciably higher proportion of these large prey items than the males.

Fig. 40. Inter-specific competition: Hen Harrier on the grouse moors.

Picozzi calculated that this predation reduced the total number of Grouse which would otherwise have been alive in August by 7.4%.

Meadow Pipits are generally the most frequent prey species, though harriers take a wide selection of the moorland nesting passerines. The smaller adult waders, especially Snipe, are often caught, and the chicks of all waders are vulnerable to the harrier's mode of hunting. There appeared to be little difference in prey spectrum between open moorland and forest nests, and Watson found that although, the passerines typical of the pre-thicket forest were taken, open moorland was the preferred hunting terrain. Males commonly ranged up to 4–5 km from their nests. Strictly woodland birds formed a very small part of the diet, and mammals (mainly Field Voles) were a barely significant item. This suggests that the attractions of young forest are not so much an enhanced food supply as an improved nesting habitat in the undisturbed and rank ground vegetation and concealing young trees. The bird's hunting style is not well adapted to performing amongst densely packed young trees, and within forests much of the hunting is along rides. Small rodents, especially Field Voles, are nevertheless evidently important food items in some localities or at certain times.

Harriers catch their prey not in spectacular aerial chase but by low, assiduous quartering of the ground, with a sudden pounce on the surprised victim. Birds are therefore safer in the air and this results in a second effect unforgivable to grouse shooters. The appearance of a Hen Harrier over a grouse moor during a shoot can cause the grouse to rise in packs and fly around wildly, so that they become impossible to drive over the guns. The day's bag of birds shot can be greatly reduced, and the keepers will no doubt often receive the blame. So precautions to obviate the risk of a ruined day's shoot will tend to be taken well in advance. Golden Eagles produce the same behaviour and are thus also unwelcome on most grouse moors. Peregrines, which usually kill in flight, tend to have the opposite effect, making the grouse reluctant to rise.

It is sad that in regard to this predator in particular, we still live with attitudes hardly different from those a hundred years ago, at least in some parts of Britain and Ireland. Until or unless the 'sportsmen' can exchange some of their pre-occupation with bag-size for the pleasure of seeing these beautiful birds sailing over the heather, the Hen Harrier will have a short expectation of life on the grouse moors. The forest nesting haunts provide a different and more encouraging story, for the bird has usually been welcomed and protected there. It is one of the more notable additions to the bird fauna produced by afforestation, though only quite locally, and it is also, unfortunately, a largely temporary bonus.

Hen Harriers are usually territorial breeders in Britain and Ireland. In many areas the nests are well scattered, with 1–2 km between nearest neighbours in south-west Scotland and Ireland. On fairly average grouse moor country in Glen Dye, Kincardineshire, Picozzi (1978) found 24 nesting territories, but only 15 definite nests, in 12 000 ha. There was a regular dispersion of nests and displaying pairs, with an average nearest neighbour distance of 1.52 km, but home ranges were large (1400 ha for one pair) with little overlap. Extremely high densities are occasionally found, evidently in response to locally prolific food supply. In 1949, 9 pairs were found nesting within an Orkney moorland valley of 250 ha, and the same number in a similar-sized area in Kintyre in 1958–60 (Watson, 1977).

Picozzi (1984) recorded 85 nests on 72 km² of Orkney moorland in 1978, but this population has been known from Eddie Balfour's work to be markedly polygynous since the 1950s, with females outnumbering males by at least two to one. Some pairings were still monogamous but many males were associated with 2 or 3 females, and Picozzi found 6 in one exceptional case. In the 1950s polygyny was correlated with a sex imbalance, favouring females, in young reared. By 1982 this had changed, for reasons unknown, to a greater proportion of males, yet polygyny had become even more marked, and the continuing relative scarcity of males seemed to stem from their higher mortality (Picozzi, 1984). While polygyny occurs elsewhere, it is infrequent and one to one pairing is usual. Newton (1979) has suggested that polygyny has selection value in raptors, in which the male typically provisions the sitting females, in situations where there is a rich food supply. Males are thereby enabled to rear more young than if they were normally paired, though females may rear fewer.

Of all our upland birds, the Hen Harrier shares with the Great Skua the reputation of being the boldest and most aggressive to man. Especially when the eggs have hatched, both sexes, but more usually the female, will sweep down at the head of a human intruder, often striking and drawing blood. Watson describes it as a thrilling spectacle and unsettling experience. The habit is evidently directed against other potential predators, and has advantage, since Watson found that nests with one or both parents aggressive were more likely to fledge young than those with timid owners. With humans it may, clearly, make the birds extremely vulnerable to destruction, since they do not discriminate when the intruder is a gun-carrying keeper.

Another unusual trait of the Hen Harrier is that of communal roosting in winter. The harriers mostly leave the immediate vicinity of their breeding places at the end of summer, and the young of the year often disperse far afield, even reaching mainland Europe. Yet, here and there, and in Scotland, usually in places adjoining the upland breeding grounds, the birds will collect together to roost. Watson (1977) followed one of these roosts in Galloway for many years. It was in a wet flow with long, tussocky grass and bog myrtle at the edge of a large sweep of moorland. The numbers of birds varied widely, reaching a peak at 30, and were influenced by weather, mild and windy nights being the best, and still, frosty ones the least productive. The geographical origin of the birds in the roost was not known. A second roost in a lowland peat moss in Wigtownshire drew two marked birds from Kincardineshire and Orkney but most of these groups seem likely to be mainly the local breeding stock. During the day the birds appear to disperse over a wide area of the surrounding country to seek food, and they may work the uplands variably, according to both food availability and weather. With the onset of the nesting season the roosts are abandoned. Their purpose is not entirely clear: certain advantages may be deduced, but if they are significant, why do other moorland predators not adopt the habit, too?

The SHORT-EARED OWL is a widespread upland breeder though it nests also in some lowland marsh and coastal habitats. A ground nester, it prefers rolling moorland to steep hillsides, and is more typical of heather, rush or cotton-grass terrain than short grassland. The Short-ear is a daylight owl and conspicuous in its ways, whether sitting on a roadside post with wide-eyed and baleful glare, beating about slowly low to the ground in search of prey, or engaging in the loftier aerial antics of courtship and territorial display. Like most birds of prey, the male feeds his mate while she is incubating or guarding small young, but when the young are bigger or fledged, she often hunts as well. On the wide sweep of moor, cold-searching for a nest after seeing a male is almost as hopeless as looking for

Fig. 41. Short-eared Owl at nest in rushes on moorland. A small brood in which the last-hatched chicks have probably already died through food shortage: possibly only the larger of the remaining two will survive. Eric Hosking.

a needle in a haystack. Watching the bird to see where he takes food can sorely test the patience, too, though it gives an insight into the problems of hunting, as well as the pleasure of observing his graceful style. A high proportion of plunges to the ground are unsuccessful, and when something is caught he will often swallow it and sit for while or just carry on quartering. Normally, the owl appears compelled to work hard to support his family, and covers a large area of up to several square kilometres in search of food.

The Short-ear is widely, though thinly, distributed over our heather moorlands. In a study of the bird on 35 km² of this habitat in North Wales, Roberts and Bowman (1986) found 4 pairs to nest regularly. Although the Pygmy Shrew was the favourite prey, in percentage weight the diet of the owls consisted of roughly equal proportions of this species, Common Shrew, Wood Mouse/Yellow-necked Mouse, Short-tailed Vole, Bank Vole and small birds. The authors believe that on heather moorland the Short-ear typically maintains fairly constant numbers and stable distributions, with nesting sites well defined and traditional, and hunting ranges large, laying small clutches at similar times annually. This matches the relative stability in overall numbers of the fairly wide prey spectrum.

In Britain, the Short-eared Owl's breeding biology is closely linked to the abundance of these rodent prey species, and the absence of all but the Wood Mouse from Ireland is one plausible reason for the bird's failure to establish as a breeder in that country (Sharrock, 1976). The Short-ear is celebrated for the marked fluctuations in its breeding populations which occur locally and periodically in response to cycles in abundance of the

Fig. 42. A key mammal of the uplands: Short-tailed Field Vole.

Short-tailed Field Vole, especially in grassy uplands. It was on the sheep-walks of the eastern Southern Uplands and Cheviots that the most remark-able increases in this owl occurred during the late nineteenth century, in response to huge irruptions of field vole populations. These 'vole plagues' occurred both on predominantly grassy hills, such as Eskdalemuir, and on heather ground. There were two main outbreaks, one during 1875–76, and a still bigger one in 1891–93.

The swarming voles rapidly drew in large numbers of Short-eared Owls, which bred at high density, beginning early (March), laying large clutches and rearing large broods (up to 10–11 young), and with usually 2 broods in the year (Adair, 1892). Then, after a year of two of peak numbers, the vole populations crashed. Deprived of this super-abundant food supply, the number of Short-ears thinned out rapidly. Some dispersed from the district but there were reports of owls found dead in emaciated condition, as though they had starved. For many years afterwards, Short-eared Owls nested mostly at rather low density over the moorlands generally, though some areas, such as Orkney, always seemed to have good numbers. There was no hint of a repetition of vole plagues until the late 1930s, when young conifer forests became extensively established in the Cheviots by the For-estry Commission. The effects on both voles and owls are such a dramatic result of afforestation that this is an appropriate place to consider further this land use change, noted as important to the Hen Harrier.

The exclusion of domestic stock from the newly afforested ground and the prevention of moor burning lead to a rapid increase in luxuriance of the existing vegetation. The grasses grow tall and dense, and dwarf shrubs such as heather and bilberry assume a new vigour of form, often expand-ing again in places where they had been reduced by heavy grazing and fire. The luxuriant grass growth provides just such an increase in both food supply and cover as the field voles need, and they often multiply in num-bers rapidly. One of the minor vole plagues occurred in newly planted for-est in the Carron Valley, Stirlingshire, in 1952–53.

Lockie (1955*a, b*) began studying the situation in 1954: although the voles had begun to decline by then, there were still 30–40 pairs of Short-eared Owls on about 1400 ha, plus 4 pairs of Kestrels. The owls were strongly territorial, and at this fairly high density (up to 3 pairs/km²) the territory limits were readily observable from the birds' behaviour, in patrolling range and interactions with neighbours. The owls hunted wholly within their territories. There was evidently much predation on their nests (probably much more than at normal breeding density), but declining vole numbers appeared to be responsible for the departure of most of the Short-ears by early June. Remaining pairs enlarged their hunt-ing territories. Lockie calculated that each adult owl consumed on average 3.06 voles every day, and each young owl needed 3.13 voles per day. He concluded that the owls hastened the decline of the voles in early spring, but when these rodents' breeding got into its stride, only a tiny fraction of their increased numbers was caught by owls. When the vole population finally collapsed, it appeared to be the result of factors other than predation. Lockie's findings support the view that numbers of predators are controlled by the numbers of their prey, and not vice versa.

Usually, the effects of afforestation are less pronounced. Certain areas have a noticeable increase of Short-eared Owls for a few years and the owl may maintain these numbers until the forest closes into thicket. The bird appears on afforested sheep-walks where it was absent previously, and sometimes shows an increase when former grouse moors are planted. Yet in some new forests it remains sparse or even absent; and even the largest vole and owl peaks associated with afforestation seem to have been less than the great nineteenth century 'explosions' described above.

The cause of these spectacular vole plagues remains a mystery, but the afforestation effect provides possible insight. While voles cycle on many sheep-walks, their peak densities are nowadays never very high. The greater abundance attained in some young forest areas seems clearly to be associated with the removal of sheep and consequent marked increase in luxuriance of vegetation, especially grasses, providing a super-abundance of food and cover for the voles. This suggests that the nineteenth century vole plagues followed local and temporary reduction or removal of sheep stocks from some uplands – perhaps in response to unfavourable economic circumstances – thereby allowing a rank flush of herbage to trigger the vole explosions, for it is scarcely conceivable that these could have occurred on ground heavily grazed by sheep.

Kestrels also increase during vole 'highs' in the young forests, provided they have enough nesting sites in older trees or rocks. Small birds find a congenial habitat in the mixture of dense ground vegetation, and the bushy young trees, which provide song posts or even nest sites for some species. Young forest is often good habitat for Stonechats, Whinchats, Willow Warblers, Meadow and Tree Pipits, and on the lower ground there may be colonisation by Grasshopper Warblers. Red Grouse persist and may even increase where heather shows recovery, and Blackgame often flourish in the young plantations. Some waders, especially Curlews, may persist for a time, but others such as Golden Plover, Lapwing and Dunlin usually drop out within a few years of planting.

After 10 years, the habitat and its bird community have changed appreciably. The trees are now asserting themselves, covering much of the ground surface with their dense growth and steadily excluding the lush ground vegetation. The moorland birds are dwindling and the woodland species increasing. By 15 years, or even earlier, the trees coalesce into a dense thicket, and ground vegetation is suppressed; it persists only along rides and roadsides, or in places which, for whatever reason, remain free of trees. The birds of open moorland have gone, totally, from the blocks of trees, and have been replaced by a group of woodland species, mainly passerine songbirds, and mostly widespread kinds at that. The once rare or very local Crossbill and Siskin have, however, become locally common in the new forests. The rides and roadside edges are too narrow to support anything more than the odd pair of Meadow Pipits, and any large unplanted areas within the forest usually have a very limited variety and number of open ground birds.

Blanket afforestation of moorlands lying below the planting limit causes a large-scale transformation of the habitat and its bird assemblages. The productive pre-thicket phase of the forest, with its often rich and varied bird community, is only temporary, and its persistence depends on the planting of yet more open ground. Nor are the effects confined to the ground directly planted. As well as the strong possibilities for increased corvid and mammal predation on nests of open moorland birds stemming from afforestation, there is some evidence that certain species, mainly waders, tend to decline on ground beyond the forest edge (Stroud *et al.*, 1987). The vegetation of adjoining ground grows taller because moor-burning is discouraged, and so the habitat becomes unsuitable for birds such as Golden Plover and Dunlin. While some displaced birds, such as Curlew, may possibly increase on unplanted upland elsewhere, this appears to be only a temporary effect and is followed by decline to normal densities in the longer term. Our understanding of carrying capacity and territorialism make it extremely unlikely that the ordinary run of upland habitats and areas have 'spare room' for absorbing bird populations displaced by large-scale land use changes elsewhere. The more probable result is that the overspill populations become subject to a new and more adverse balance between recruitment and mortality, which leads to their demise within a few years.

The eventual clear-felling of forest blocks at 30+ to 60 years will recreate a patchwork of open ground conditions at any one time. We are already able to observe what happens on such clear-fells in older areas of upland forest, such as Kielder and parts of Galloway. Their vegetation does not revert to that of the open moorland before afforestation. A dense grass-dominated community usually springs up, with a patchy development of heather and bilberry in places. The 'slash' of dead branches from the felled trees is usually left, and adds to the unevenness of the ground, which retains the ridges from earlier ploughing. A few open ground species, such as Red Grouse, Curlew, Nightjar and Blackgame may typically recolonise, but birds of woodland edge and scrub or glade predominate. The simple list of species which are recorded nesting within open forest habitats or clear-fells is not valid evidence that afforestation retains much of the previous ornithological interest. It is the frequency and overall numbers of open ground species which matter, and upland afforestation has demonstrably caused and is causing a huge and continuing decline in the populations of upland birds.

The impact of afforestation on birds and their habitat has been described in several recent papers (NCC, 1986; Ratcliffe, 1986; Stroud *et al.*, 1987; Thompson, Stroud and Pienkowski 1988). This is one of the most serious of the conservation problems facing upland birds and is considered further in Chapter 9.

4

The deer forests

There are Red Deer on Exmoor, and Martindale in the Lake District is an English upland deer forest from which this animal has spread to other fell groups. The Galloway hills have a fair population, and in Ireland, there are Red Deer in the mountains of Donegal and Kerry. The main deer forest country is, however, in the Scottish Highlands, and especially the higher, remoter parts. Red Deer stalking became a fashionable Victorian sport, and much of the Highlands was parcelled out into estates whose primary purpose was the pursuit of this, our largest surviving land mammal, with the rifle during the autumn. Wealth derived from the Industrial Revolution and from the Empire flowed freely into this previously remote and neglected region. And so was founded the vogue of the shooting lodge, of varying architectural grandeur, with it mostly absentee owner, and locally resident factor, deer stalkers and gillies. A minority of estates were managed for both deer and grouse.

Some deer forests were on relatively fertile mountains, such as the calcareous mica–schists of Glen Lyon, Caenlochan and Clova, all happy hunting grounds for the Victorian botanists. Yet many famous forests were on poor rocks, such as Black Mount in Argyll, Mar and Lochnagar on Deeside, Rothiemurchus and Ardverikie in Inverness-side, and many of those north of the Great Glen. The Red Deer has a notable ability to live on poor ground, including our most sterile mountains. The animals are mostly small compared with European forest Red Deer and low carrying capacity restricts their performance on many forests. Yet they can manage reasonably well on their own under the bleak climate on ground where sheep need a modicum of winter feeding, besides the usual chemical attentions to relieve them of a medley of parasites and pathogens. Sheep are nowadays run on many of the deer forests, though at varying stocking rates.

Many deer forests have little woodland, and some are virtually treeless. Most of the original woodland had been destroyed by 1800, before interest in deer arose; it was cleared for timber and charcoal, to create grazing range for domestic stock, and to eradicate shelter for wolves and perhaps for human enemies. When Harvie-Brown speaks in his Fauna series of 'afforestation' he is referring to the creation of deer preserves. The real sort was to come much later. While existing woodland gives good winter shelter to Red Deer, this animal has helped to keep the Highlands in a deforested condition. It is a most formidable suppressor of tree regeneration, by cropping seedlings and barking saplings before they have a chance to form trees. The beneficial effects of excluding deer by high fences are startling: wherever parent trees are around, seedlings of Scots pine and birch usually come up thick and fast, while there are few or none outside. And unless measures such as this are taken, or a drastic culling of the deer carried out, many Highland woods are effectively moribund, for when the present trees die, none will have established to take their place.

Red Deer live through much of the Highlands at unnaturally high density. There are no significant wild predators of the species in Britain now, and while the annual cull of stags and hinds maintains numbers within prescribed limits, most estates manage for an optimum stock of animals. There is usually a fairly high natural mortality, which gives a certain amount of carrion. Where sheep are present they add to the supply. And there are the wild animals of the hills: Mountain Hares, Rabbits, Red Grouse, Ptarmigan and Blackgame as year-round residents, plus the seasonal visitors among the birds.

Fig. 43. A pair of Golden Eagles at their eyrie on a cliff ledge. The male brings prey to his mate but both share incubation duties. Dennis Green.

Between them, these mammals and birds provide food for the GOLDEN EAGLE, the most spectacular of all our upland birds, which has its traditional home and sanctuary in the Highland deer forests. With the Red Deer itself, the Golden Eagle is the grandest of the wild creatures which fittingly belong to this most extensive and challenging of our mountain regions. Massive and majestic at close quarters, its high-soaring form is similarly inspiring against a background of dark pinewoods, lofty crags and snowy corries in its classic haunts amongst our highest mountains. The name of Seton Gordon will forever be linked with the Golden Eagle in the Highlands. His writings during the early decades of this century vividly portray the Cairngorms and other mountain ranges as remote and lonely country, accessible only to the more determined walkers. The mantle of Golden Eagle watcher-in-chief has passed to Adam Watson, whose work has given

Fig. 44. Predator and prey: Golden Eagle after Ptarmigan.

modern insight on the ecology of this bird, though other eagle enthusiasts have also contributed important studies.

Seton Gordon and others regarded the deer forests as the Eagle's salvation, as areas where Grouse were not preserved, or were even unwelcome, from their way of spoiling the stalking by rising in noisy alarm (Gordon, 1955). For this is a predator which fires the imagination in quite different ways wherever the sporting interest in deer is exchanged for that in Grouse. And in sheep country the bird's unpopularity is still registered as periodic complaints over lamb-killing.

As far as its history is known, this has always been an upland bird in Britain and Ireland. Golden Eagles are believed to have occurred in the eighteenth century as far south as Snowdonia (known in Welsh as Eryri, the land of eagles), the Pennines, Lakeland and the Cheviots. In the Southern Uplands they survived in Galloway and the Moffat–Tweedsmuir Hills until the late nineteenth century. The species bred widely in virtually all the main mountain groups in Ireland. Ussher and Warren (1900) mentioned breeding haunts in Waterford, Tipperary, Kerry (numerous), Galway, Mayo, Sligo, Leitrim, Donegal, Londonderry or Tyrone, Antrim, Down and, less certainly, Wicklow. At the time they wrote, only a few pairs remained in the remotest regions, and disappearance was complete by 1912. All this retreat was blamed upon deliberate destruction by gamekeepers and shepherds, with collectors helping to give the *coup-de-grâce*. Leslie Brown (1976) has discussed at length the treatment that this and other birds of prey have received at the hands of Man.

The Golden Eagle's fortunes have improved variably in recent years. Seton Gordon believed that numbers increased in the western Highlands during 1900–50. The two war periods 1914–18 and 1939–45 may have helped, by reducing gamekeeper activities. The species has certainly reappeared in some old haunts from which it had vanished. Perhaps it was never truly absent from Galloway and Carrick, but regular nesting was only known for certain from 1945 onwards, building up from 1 pair to 4 pairs by 1963. Sporadic nesting has also occurred farther east in the Southern Uplands since 1972. Golden Eagles did not certainly breed again in Lakeland until 1969. A second pair began to breed in 1977, but dropped out after 1983. From 1976 onwards a third pair has bred elsewhere in Northern England. Wandering Eagles are seen frequently in northern England nowadays, and would most probably settle to breed if left unmolested. They are rarely seen in Wales, however, and there have been no reports of attempted breeding. There was a brief return to Ireland, when a pair bred on the coast of Co. Antrim from 1953 to 1960, but it was evidently merely an alternative haunt for a pair on the Mull of Kintyre 23 km away, and the birds returned to Scotland. Ireland otherwise appears to have remained without Eagles, which is surprising, for there is so much suitable country still, and no reason why some juveniles should not head that way.

The Golden Eagle is a shy, silent and somewhat elusive bird, for all its size. It can be around for some time before coming to the notice of casual observers. While Buzzards are often reported as Eagles, the reverse also happens, and a high soaring bird may escape recognition. Seen against a hillside, an adult is much darker than a Buzzard, and the Gaelic name of Iolaire Dubh (Black Eagle) is at once understandable. Sometimes an Eagle can be watched close to a road, but it is possible to travel far through Eagle country before seeing even one bird. Incubating Eagles can sit extremely 'tight' so that an occupied eyrie may easily be missed. When the sitter is flushed it usually flies straight away without any fuss, to circle silently or seek a distant perch, and await the departure of intruders before venturing near the site again.

Golden Eagle country varies in character. First is the nearly treeless mountain country so typical of the western Highlands, and, formerly, of the Galloway hills. Much of it is rugged and some of it extremely so. Many of the nesting places are in big cliffs, sometimes in high corries where snow lies late, or on flanking escarpments far up the main glens. Stream gorges are quite often used, and sometimes hidden ravines on steep hillsides. In less rugged areas the birds will often be satisfied with some small and inconspicuous outcrop, and there are eyries where the tale could be true, that the shepherd's wife walked in and carried off the eggs in her apron. Another type of terrain is the well wooded slopes of the Cairngorms–Lochnagar massifs, and the glens flowing to the Beauly Firth, where some Golden Eagles nest in trees, usually Scots pines, but occasionally birches. The nest trees are nearly always in very open woodland, where pines are scattered individually or as clumps in a heather moorland, so that the birds usually have a good outlook and can hunt

Fig. 45. World breeding distribution of Golden Eagles (from Voous, 1960).

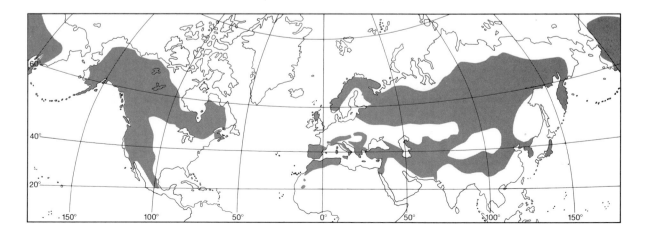

amongst the trees. In the same region are mainly heathery hills with a few trees and varying amounts of crag, but good populations of wild prey. Eagles also nest in seacliffs on the west and north coasts of the Highlands and in the Hebrides. These coastal haunts are nearly all backed by continuous sweeps of moorland, extending from the cliff tops inland over areas sufficient to provide hunting range. Some of these coastal Eagles are in stupendous precipices, such as the basalt headlands of western Skye.

Golden Eagles are amongst our hardiest birds. They cling to the mountains during the hardest of winter weather, though heavy snow may drive them to the lower fringes for a time. The settled adult territory-holders appear to be highly sedentary, roosting either on the nesting cliffs or in particularly favoured places elsewhere. The eyries are often patched up or built anew in the autumn, so that when nesting activities begin in earnest, during the following March, lining the nest is the main task left. New nests are often small and inconspicuous, but old and much used eyries can become great heaps of sticks and heather clumps, visible from afar. The lining in most usually of great woodrush but the leaves of other grasses such as *Molinia* are sometimes used. Tree eyries are sometimes the largest of all, for it seems more usual for these to be used year after year. The most famous, in the open fringe of a Deeside pinewood, reached a height of 5 m.

Most pairs of Golden Eagles adopt the raptor habit of using alternative nest sites in different years. On many cliffs there will be, at any one time, 3–5 eyries of graded age; their order of use varies between different localities, but is often irregular. The number of nest sites known to be used tends to increase with observation period, and in territories followed over many years, up to 10 different nest sites are not uncommonly known. In rugged country with numerous cliffs, the alternative sites are often spread between different crags, up to 5 km apart. This use of alternative cliffs within a territory complicates eyrie-checking, and can lead to some occupied nests being missed. In especially precipitous terrain it may take two days or more to search even a single te itory. Old nest sites often become conspicuous through the greenness of the ledges and rocks below, as a result of the nutrient enriching effects of the excrement and prey remains. Mosses, lichens and algae often grow in particular abundance on the faces immediately beneath, while growths of woodrush, other herbs and ferns on the ledge itself are frequently especially profuse.

This is a slow-breeding species. Golden Eagles do not attain adult plumage until 3.5 years and, while immature birds are sometimes paired with an adult in a nesting territory, breeding does not usually occur until age 4 years. Although two young are usually hatched, asynchronous hatching in this bird is reinforced by the innate tendency of the first hatched chick to attack its younger and weaker sibling, the 'Cain and Abel' struggle often resulting in the death of the latter. In the Western Highlands, Galloway and northern England, the great majority of successful nests produce only a single youngster. Broods of two fledged young occur mostly in the eastern Highlands, where live wild prey is most abundant. Exceptionally, three young have been fledged in a single brood.

While there is little direct evidence about age distribution in our Golden Eagle population, birds with these breeding characteristics are usually long-lived, and have a good life expectancy when protected. Equally, when persecution causes a high mortality, they are vulnerable to rapid population decline. In the main deer forest strongholds, established breeders probably have a good survival rate. Some of the immature birds, recognisable by the white bands across the base of their tail and white wing patches, remain in the same areas, and represent a non-breeding surplus. They will supply replacements for breeders which die, though if they pair before becoming adult it may be one or more seasons before they begin to breed themselves. Some youngsters also wander away from their birthplace, no doubt impelled by an urge to seek a place of their own. Many of them reach the grouse moors of eastern or southern Scotland, or northern England, there to run the gauntlet of gun, trap and poison. Some probably seek the foothill country and lower sheep-walks where the odds are also loaded against survival. Here and there birds last long enough to establish a new breeding territory once in a while. But the majority sooner or later meet a premature end. A good many Golden Eagles have made their way into the Cheviots and Pennines during the past 30 years, but only in one place, to my knowledge, has nesting ever occurred. The pair have had strict protection here.

Golden Eagles are territorial birds, though the territories appear usually to be established through more subtle kinds of interaction than actual aggression (Brown and Watson, 1964). This includes aerial display by one or both of the pair, either by high circling and patrolling or by the undulating 'sky-dance' flight, in which a bird repeatedly dives and then climbs back to its previous pitch. Fights with intruders or neighbours are rare, and other Eagles appear to respond to territorial display by an avoidance reaction. Doubtless the threat of combat lies behind this behaviour, and will develop into actual fighting when any intruder ignores the visual signals. The habit of using widely dispersed perches and roosts within the territory, as well as alternative nesting cliffs, probably helps to establish other visible evidence of possession and presence. Buzzards are known to place green sprays of foliage on ledges in addition to the actual nest sites as a form of territorial marking (Fryer, 1986). Yet Brown and Watson (1964) believed that neighbouring pairs of Eagles often overlapped in their hunting ranges.

This territorial behaviour maintains a regular dispersion of breeding pairs and stability of breeding population. Average territory size and, hence, breeding density, vary within the Highlands. As with other crag-nesting species, an obvious limitation is placed upon breeding density when suitable crags are in short supply, as they are in some of the gentler moorland areas. Even though Golden Eagles will adopt quite insecure sites on tiny rocks, there are limits to their adaptability in areas such as the Flow Country. They appear to be more particular over choice of tree sites, in that many areas apparently suitable, both as regards hunting habitat and presence of woodland or trees, are not occupied by nesting Eagles. A few pairs nest, in different years, in both rocks and trees, and artificial shelter woods are occasionally used.

In many parts of the Highlands, nest site availability is not limiting, how-ever, and breeding numbers appear to be at a saturation level. This varies somewhat, and a recent study by Watson, Langslow and Rae (1987) has shown that variations in nesting density are closely correlated with amounts of winter food, especially carrion. Areas of high density, with more than 25 pairs/1000 km², are mainly in the west. Breeding densities are generally lower in the east (fewer than 15 pairs/1000 km²), but breed-ing performance higher.

Sheep farmers were formerly major persecutors of Golden Eagles, and in some parts of the World they probably remain so. In this country, there is increased acceptance that most of the mutton and lamb eaten by eagles was already carrion, and though many shepherds remain suspicious and unwelcoming, probably most of them respect its status as a specially pro-tected raptor. Complaints of lamb-killing on Lewis, in West Ross and west Inverness have all concluded that, while this predation undoubtedly occurs, it usually accounts for a very small proportion of live lambs. The most recent investigation, at Glenelg, calculated that Eagles killed 1.8% or less of the lambs born in the area, while overall mortality was 26% (Leitch, 1986). Here, as elsewhere, individual birds can develop a predilection for taking live lambs. Lockie (1964) found that in Wester Ross Eagles ate less lamb in years of good lambing than in years when lamb mortality was high, and did not make up the deficiency by increased lamb-killing. While this is a local or periodic problem, the overall loss of live lambs to Golden Eagles thus seems insignificant.

The birds's fondness for sheep carrion began to put it in difficulties of a quite different kind. In the late 1950s and early 1960s, Golden Eagles in the Western Highlands were clearly showing some of the same symptoms that were closely associated with population decline in the Peregrine and Sparrowhawk. Many eyries had broken eggs, or were found to fail through desertion, addling of eggs, and death or disappearance of the young. Although in this species it is not unusual for established pairs to fail to produce eggs, non-laying after the building up of eyries also became more frequent. A sample of 39 eyries in western Scotland during the period 1963–65 showed only 31% breeding success, compared with an earlier sample of 40 western eyries during 1937–60 which had 72% success (Lockie and Ratcliffe, 1964). Chemical analysis of eggs showed that the adults were consistently contaminated by residues of DDT, lindane and dieldrin which appeared to originate from sheep dips incorporating these chemicals. Analysis of castings showed that this western population of Eagles was taking a high proportion of sheep and lamb carrion in its diet.

While these organochlorine residues were rather small, there appeared to be enough DDT and its breakdown product DDE to cause the 10% decrease in shell thickness found in a sample of eggs. The combined pollutant levels also seemed sufficient to account for the other adverse effects on reproduction which had been noted. In early 1966 the organochlorine compounds were withdrawn from sheep dips, largely through concern about hazards to humans eating contaminated mutton. There was a quite dramatic recovery in Golden Eagle breeding success, and in a further sample of 45 eyries examined during 1966–68 in the same western region, 69% were successful in fledging young (Lockie, Ratcliffe and Balharry, 1969). Mean organochlorine residues in samples of Eagles' eggs were halved during 1966–68, compared with 1963–65. During the whole period 1963–68, a closely studied eastern Eagle population maintained a high breeding success of 70–80%, and organochlorine residues in eggs remained very low. Although there was no convincing evidence for a decline in Eagle population, and the sheep dip effect was on breeding performance, the depression in output of young birds must, if it had continued, eventually have caused the number of breeders to decrease in the west of Scotland.

In Lakeland, two single adult Golden Eagles were found during 1957 in possession of eyries in widely separated localities. In each succeeding spring, both birds built up their nests, but remained apparently unmated, and so breeding could not occur. Hopes were raised when two adults were seen near one eyrie in September, 1958, but the following spring only one Eagle was in evidence again. While these slow breeding birds were known to take several years to establish themselves in full breeding in a new or long-deserted locality, this protracted failure to reproduce pointed to some particular adversity. Other eagles appeared in the district so that lack of potential mates seemed unlikely. Inhibitory effects of the organochlorine sheep dips seemed possible. Sure enough, soon after these were withdrawn, one bird paired, and the first authenticated eggs of Golden Eagle in the Lake District were laid in 1969. A pair has bred in this locality in every succeeding year, usually rearing a single youngster, under careful protection by the RSPB.

The pesticide threat to the Golden Eagle passed, for the live wild prey occurs in a largely uncontaminated environment. By the 1970s, the total population was probably as large and stable as it had been for a long time. Since then, new problems have begun to appear. Numbers in some of the eastern Highland areas studied by Adam Watson have decreased, and tree nesters have declined in particular. Some of the nesting places are much disturbed by walkers and climbers, and this may have caused nesting failures or even desertions. Still more important, however, is the increase culling of deer stocks in some areas, which has reduced mortality and, hence, the amount of winter carrion available to Eagles (Watson, Payne and Rae, 1989). Since sheep are few in these areas, the food supply has decreased appreciably and territories have been lost.

The other probably more serious problem is that of afforestation. Golden Eagles can flourish in partly wooded country and extensive areas of open 'park woodland'. They are, however, unable to hunt within closed forest and so cannot cope with the great blankets of dense plantation which modern tree-farming forestry so usually creates. In Galloway and Carrick, four pairs of Eagles were breeding regularly, and with reasonable success overall, by 1966. Afforestation has expanded rapidly in this district, and encroached increasingly upon the hunting ranges of these Eagles. The breeding performance of this small population has, since the early 1970s, shown a decline which coincides closely with the scale and pattern of this encroachment. One pair ceased breeding altogether and has now disappeared (Marquiss, Ratcliffe and Roxburgh, 1985). Watson et al. (1987) have found evidence of decline in heavily afforested parts of the Highlands, and predict that this could cause an overall loss of 20% or more of breeding population by AD 2015, even if no more land is planted (see p. 218).

Despite various adversities, especially persecution, the Golden Eagle has been one of our more successful predators, and its estimated breeding population of at least 424 pairs in 1982–83 (Dennis et al., 1984) is one of the healthiest in Europe. There is still some mystery in the contrast between this species and the SEA EAGLE, which was once a widespread breeder in Britain and Ireland, but had declined to extinction by 1916. The Sea Eagle's demise is attributed largely to destruction by sheep rearers and game preservers, with assistance from skin and egg collectors in the later years. Yet the Golden Eagle suffered this treatment, too. Whether the Sea Eagle had some additional ecological or competitive disadvantage, or greater vulnerability to destruction, is not known. Although towards the end, the Sea Eagle became restricted to coastal cliff haunts in the Highlands and Islands, and western Ireland, it was undoubtedly a widespread bird of the mountains as well in earlier times.

The recorded history of eagles in the Lake District refers almost entirely to Sea Eagles, though Golden Eagles were evidently present too. The celebrated eyries periodically raided by sheep farmers were all those of the Erne, at Eagle Crag in Borrowdale; Eagle Crag near Buttermere; Buck Crag in Martindale; and Wallow Crag above Haweswater (Macpherson, 1892). There are in the district at least a dozen Erne (or the derivative Iron or Heron) Crags, suggesting a fair population of the species. In Scotland, Sea Eagles sometimes nested on small islets in mountain lochs, such as L. Skene in the Moffat hills, the Fionn Loch in Wester Ross and L. Fiag in Sutherland. Several inland cliff nesting haunts were mentioned in the north-west Highlands, including the great precipice of Beinn Airidh Charr in the Fisherfield Forest (Harvie-Brown and Macpherson, 1904).

The Sea Eagle seemed so unlikely to recolonise Britain on its own that deliberate reintroduction eventually became agreed as a desirable move. Three birds were released in Argyll in 1959 and four on Fair Isle in 1966, but numbers were too small for success and they disappeared. After careful discussion between the interested parties, the NCC in 1975 launched a longer-term project for reintroduction, based on the island nature reserve of Rhum, where the birds could at least have sanctuary at the start. Ian

Fig. 46. Return of a lost native: Sea Eagle in a mountain corrie.

Newton gave technical advice on the reintroduction programme. With the cooperation of the Norwegian Government, Norwegian ornithologists (notably Harald Misund), the Royal Norwegian Air Force and the Royal Air Force, a total of 82 young Sea Eagles was transported from northern Norway to Rhum from 1975 to 85, to be reared to fledging and then released.

The Project Manager, John Love, responsible for the care and release of the Rhum Sea Eagles, has provided an interesting account of the venture and its success, at least in meeting the first goal, of restoring this species as a breeding bird to Britain (Love, 1988). Seven of the introduced birds have been found dead, but there appears so far to have been good survival among the rest. Attempted breeding first took place in 1983, and the first young Sea Eagle was reared in 1985. Seven pairs were located in 1986, five of them attempting to breed, and the successful pair of the previous year reared two chicks. In 1989, 12 pairs held territory, 7 attempted to breed and 3 reared a total of 5 young (NCC/RSPB).

Ultimate success will be the establishment of a self-maintaining breeding population of Sea Eagles. The portents seem hopeful. From the release sanctuary on Rhum, where the species nested up to 1909, the birds have dispersed widely through their former range in the Hebrides and West Highland mainland, occupying territories on both rugged coasts and mountainous deer forest. Nesting has occurred in trees, as well as on both coastal and inland cliffs. The birds mostly appear to find an adequate food supply, and the earlier fears of sheep farmers and game preservers have largely been allayed. There has been general goodwill and cooperation from local people over the venture, and it has been widely received as a positive piece of wildlife conservation. The project has involved the efforts of a large number of people, and has been a team effort, involving collaboration between the NCC, ITE, RSPB, World Wide Fund for Nature, Scottish Wildlife Trust, Eagle Star Insurance and Britoil.

The Golden Eagle may be better adapted to living in barren mountain country than some of the other birds of prey. There is some evidence that the northern and western Highlands, where it is most numerous, suffered a run-down in soil fertility and carrying capacity for animals as a result of long-continued and extractive land-use, involving repeated removal of animal crops and moor burning. This is believed to have caused permanent nutrient depletion and loss of biological activity in soils, with increased acidity and declining nutritive value of vegetation (McVean and Lockie, 1969). The most quoted evidence of such effects is that of the Highland laird Osgood MacKenzie (1924) who observed substantial declines of Mountain Hares, Red Grouse, Ptarmigan, Black Grouse, Golden Plover, Snipe, Lapwing, Dunlin and Partridge in Wester Ross during the period 1860–1900. Harvie-Brown and Macpherson (1904) also commented on the deterioration of many once productive grouse moors in the western Highlands and Islands, and Grouse are now mostly sparse throughout this region. Golden Plover also appear to have shown general decline. A similar run-down in carrying capacity and moorland bird populations probably occurred in Western Ireland (John Wilson).

Golden Eagles now occupy a good many cliffs known as regular Peregrine haunts in this region during 1860–1900, and the one may have increased as the other declined. Peregrines and Ravens are mostly sparsely distributed in deer forest country, and the Eagle is often more numerous than either. Peregrines could be expected to decline through decrease in their main prey species.

Over-stocking of some of these barren uplands with sheep and, perhaps, Red Deer, has produced an abundance of carrion to explain the increase in Eagle numbers. There could be competition effects as well, since the Golden Eagle's diet overlaps with those of both Peregrine and Raven. The last two species seem to avoid nesting close to Eagles, and will vacate their regular cliffs if Eagles move in. Both adult and young Ravens are sometimes taken as prey by Eagles and, although Peregrines appear to have the upper hand in the air, they and their young might be vulnerable to predation at the eyrie.

Buzzards are also rather thinly though widely scattered through the deer forests. In some lower or coastal, and well wooded parts of the western Highlands and Islands they are fairly numerous, as in Argyll and on Mull. Merlins are somewhat sparsely distributed though perhaps often overlooked, and Hen Harriers are absent from many of the more barren areas. The most common predator is the Hooded Crow, but this bird is also more numerous in the more fertile straths and along the coastal fringe. Red Grouse are patchy but usually at low density, and in some deer forests one can walk a long way before seeing a bird. Golden Plover are mostly sparse, and Dunlin found mainly where there are patches of flow bog. Even Meadow Pipits are often in low numbers. On some of the deer forests on the most acidic rocks, total bird populations may not exceed an average of 10–15 pairs/km^2.

The birds of lake and stream (Chapter 2) are often well represented, and some of the waterfowl have significant population segments in the deer forests; both widespread species such as Goosander and Red-breasted Merganser, and local or rare ones such as Wigeon, Greylag Goose and Common Scoter. The Common Gull is widespread on the lochs of the deer forests, usually in small colonies from a few to 50 pairs, but occasionally as solitary pairs. It favours lochs with islands or marginal sedge swamp giving greater security from mammal predators, but also nests on spongy flow ground and drier moorland. It is a high-nesting species, breeding up to 1000 m in the Cairngorms. This chapter will end with an account of two birds which both belong especially to the wild and inhospitable lower moorlands of the northern deer forests, and depend on open water habitats.

The most special bird of the deer forests is the GREENSHANK which, apart from recent and isolated nesting in western Ireland, breeds only in the Highlands and some of the Western Isles. Though widespread it is mostly thinly scattered and its great stronghold is in Sutherland, Caithness and Ross-shire. We owe most of our knowledge of the Greenshank as a British nesting bird to the energies and dedication of Desmond Nethersole-Thompson and his family. As a young man, Desmond fell under the spell of the Highlands and their birds, and devoted much of his life thereafter to unravelling the secrets of breeding biology of the least known species. The Greenshank became a special challenge, from the difficulties in the finding of its nests; and the fruits of his patient studies are available to us in two notable monographs. The earlier New Naturalist *The Greenshank* (1951) was based especially on study of the small Spey Valley population where this captivating bird nested in boggy or heathery clearings and open moorland among the beautiful pine forests of the Cairngorm foothills (Fig. 5). *Greenshanks* (1979) represented a great expansion of experience, drawn especially from 15 years of annual family expeditions to a remote Sutherland glen, where the bird was in contrasting, treeless terrain.

Greenshanks mostly return to their Scottish breeding haunts during early April, and for 2–3 weeks afterwards are conspicuous and vocal in pairs as they court and stake out territories. It is then that the male gives his thrilling song-dance in undulating flight high over his chosen area, advertising his possession to all his neighbours. This is a water's edge feeder, probing and picking for the wide range of invertebrates along the margins

Fig. 47. World breeding distribution of Greenshanks (redrawn from Voous, 1960).

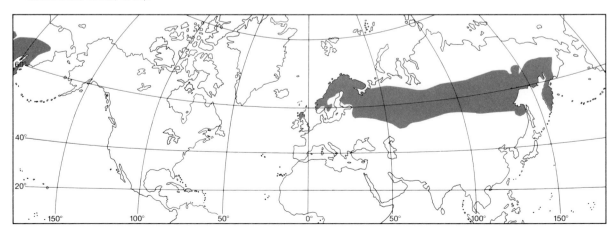

of lochs, lochans, bog pools and rivers or on the seashore. Favourite feeding places are moorland lochs with sandy and gravelly margins, giving way in places to sparse fringing growths of bottle sedge or other emergent plants typical of nutrient-poor waters. In years when April is cold and invertebrates are less active early on, Greenshanks nest later (Thompson *et al.*, 1986).

By early May, Greenshanks are to be found feeding singly. They rise to

Fig. 48. Breeding distribution of Greenshanks in Britain and Ireland during 1968–72 (from Sharrock, 1976).

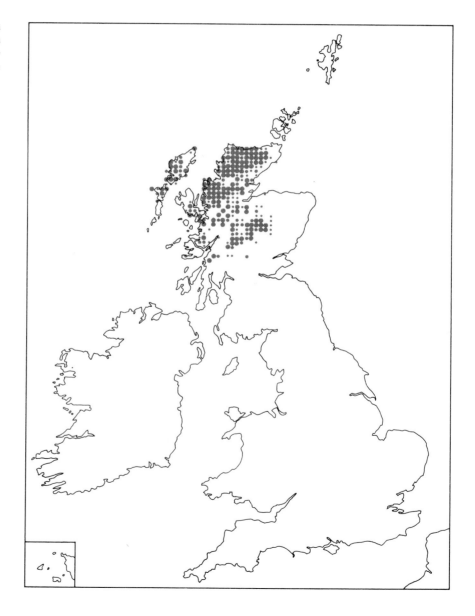

the intruder with an alarmed 'tew-tew-tew' and rapidly distance themselves before pitching to resume their quiet search for food. The absent mate is sitting on eggs in some secret place on the moor: it may be near or far, but there is not the least clue to suggest where. The sitter may fly if a human passes within 30 m or so, but many will stay until almost trodden upon, especially towards the end of incubation. Occasional nests are found by chance in walking over the moor, but cold searching is seldom worth attempting. The chink in the Greenshank's armour is that twice in every twenty-four hours, usually, the off-duty bird will fly from its feeding place and change over with its incubating mate, which then takes its turn to feed by water. The art of the game is to sit patiently at a well-chosen vantage point to watch the feeding bird and track its eventual flight to the nest.

This is much easier said than done. The non-sitter may move between a widely spaced selection of feeding places during the day. The nest can be as close as a few metres from the water's edge, or it may be at least 8 km from the main feeding ground. The frequent wind, rain and mist of the Highland spring often defeat attempts at nest-tracking. One of the mysteries is that the off-duty bird knows precisely where the nest is and heads unerringly for it, however featureless the sweep of moor may appear to the human eye. Such a finely tuned homing ability is a necessity in such birds, which nest on the ground but feed often far away and out of sight or sound from their mates.

Fig. 49. Greenshank at the nest on stony moraines in North Uist, Outer Hebrides. Both sexes take turns at incubation and their change-over is the only clue to the whereabouts of the nest. Dennis Green.

In the west Highlands the nest is often on dry stony moraines with sparse growth of heather, deer sedge, flying bent, mosses and lichens. Sometimes it is in shallow bog or wet heath, with denser though short growth of these plants. On the great flows of the far north it is often on a drier hummock amidst wet bog, sometimes surrounded by systems of pools. A large proportion of nests are placed against a stone, but in open forest habitats a piece of tree stump or fallen timber is typically used instead. Nests have also been found against broken pieces of snow fence. The mottled grey back of the sitting bird admirably matches the lichen growths on such objects, or dry patches of fringe moss, and the effect doubtless gives extra protection against the searching gaze of predators.

Behaviour changes as soon as the young have hatched. Both parents are now commonly in attendance, and begin careering around the moor with frenzied alarm calls, even when humans are several hundred metres away. This performance becomes intensified as the chicks are approached, and one or both parents often dash down close to the intruder, or pitch nearby to yelp away in protest. The chicks sometimes give themselves away by calling back, but the adult can silence them with a special call. They are usually taken to water soon after hatching, and the brood can sometimes travel a remarkable distance from the nest in a short time. Many broods are moved 3 km or more from where they are hatched, and this can complicate the recording of precise breeding distribution and numbers. The parents have even been seen to entice the young to swim across quite large rivers. It would seem simpler for them to carry the chicks in Woodcock fashion, but this has only been observed once, in Sweden. The dubh lochans of the northern flows appear to be locally important feeding places for Greenshank families, but the shores of larger lochs are the more usual places where the young fledge, by this time with only the male in attendance.

The Greenshank thus contrasts with moorland waders such as the Golden Plover, which makes very little use of open water habitats for feeding. The Common Sandpiper also depends largely on lake edges and rivers and their vicinity for food, but seldom nests far from water. Throughout most of their range (the Subarctic–Boreal zone of Eurasia) Greenshanks appear to be mainly birds of open heath or bogs with sparse tree cover, but within the taiga forests. In Scotland they have adapted to largely treeless habitats within submontane ground which was once forest-covered, though the wetter flows are likely to have been devoid of trees for thousands of years. The majority nest below 450 m, on ground which is often uneven and sometimes rocky, but with only a low or negligible overall angle of slope.

Part of the difficulty in studying the Greenshank in Scotland is that, at best, it is a sparsely distributed bird. Of the more widely distributed waders here, it is the rarest, perhaps because nearly all its breeding haunts are on sterile moorland with low carrying capacity for both terrestrial and aquatic animal life. While single pairs nest in some of the lower corries, this is not a bird of steep or high places, so that it tends to be thinly and irregularly scattered among the lofty mountain ranges covering so much of the Highlands. It is more continuously distributed on some of the lower and gentler moorlands, but the best areas are mostly where there is a good density of small to medium sized lochs, or large rivers. A pair to 1 km^2 on suitable terrain is a good density but may be doubled in especially favoured areas. The Sutherland population still being studied by the Nethersole-Thompsons has varied between 6 and 23 breeding pairs, with an average of 14 pairs during the 25 years 1964–88, over an area of *c.* 21 km^2. Fluctuations appeared to depend partly on breeding success of the previous year. Nest spacing varied, 5 valley bottom nests averaging 1200 m apart, whereas 5 on a higher and more open moorland part of the area averaged only 650 m apart. The valley bottom birds also nested earlier, but did not breed more successfully than the others.

This appears to be consistently one of the densest Scottish populations, on average. Measurements of breeding density in this species in some areas reflect the limitations of feeding habitat availability rather than true territorial spacing. Dry eastern grouse moors with few feeding lochs mostly have few and widely scattered Greenshanks, but coastal feeding places sometimes compensate for lack of inland waters, or are used as alternatives. Average breeding density in the Flow Country of Caithness and Sutherland is not particularly high, but the total area is so extensive that it contains the largest segment of the whole British population. From measurements of the total areas of different habitats and the densities associated with these, Stroud *et al.* (1987) calculated that the Flow Country held a population of 630 pairs. The total numbers in Scotland evidently approach 1600 pairs (Stroud *et al.*, 1990).

It is surprising that Greenshanks have never been found nesting on the wild moorlands of Galloway and Carrick, a country of lochs, moraines and flows extremely similar in appearance and vegetation to the typical Greenshank ground of the western Highlands. I was always hopeful of encountering a pair during my wanderings there, but never had such good luck. It seems that the species evidently responds to conditions of climate or day length which impel it to travel farther northwards, beyond the Highland line, in seeking its spring quarters. Yet one pair, at least, nested for several years much farther south, in the Mayo–Galway region of western Ireland (Sharrock, 1976) and the recent trend towards colder springs might have been presumed to favour a southward spread. The evidence for marked and long-continued acidification of many moorland lochs in Galloway raises thoughts about the possibly adverse effects of this post-

Industrial Revolution change on birds such as the Greenshank. The western Highlands are farther from major sources of atmospheric pollution, so that acidification of lochs and streams there is less marked.

The Greenshank was for long one of the more secure of our rarer birds. Egg collectors were always drawn by the elusiveness of the nests and the beauty of the eggs, but the number of clutches taken in any one year must always have been insignificant. Most Greenshanks bred in virtual wilderness country where few nests were ever seen by human eye, and though other predators took an annual toll, deer forest management and low fishing intensity seem to have suited them very well. Major hydro-electric schemes which created widely fluctuating water levels around some of the larger lochs may have spoiled some feeding grounds, though this has not been studied. Yet, within normal fluctuations of the kind recorded by the Nethersole-Thompsons (1986) and evidently resulting partly from adverse weather years, numbers seem likely to have remained remarkably steady over the first half of this century.

Sadly, this has all changed within the past 20 years. Expanding afforestation in western and northern Scotland has made rapid inroads into the Greenshanks' country, obliterating their nesting grounds and disturbing the rivers and lochs where they feed. Stroud *et al.* (1987) estimated that at least 130 pairs of Greenshanks had already been displaced by afforestation on the Caithness and Sutherland flows. The portents are for still further losses, and since Greenshanks nest almost entirely below the potential tree-line, the species is greatly at risk to this profound land use change. The enhanced acidification of stream and lake waters associated with afforestation (NCC, 1986) may further depress the carrying capacity of affected uplands for this bird.

The small population in the Cairngorm foothills is now greatly reduced, with the forest nesters almost gone (Nethersole-Thompson and Watson,

Fig. 50. A Highland fisherman: Black-throated Diver on a hill loch.

1981). The Speyside groups depended on large clearings created by forest fires and logging, as well as on naturally open bogs. Decline began in the 1940s, as the heather grew long and small feeding places dried out. The direct contribution of afforestation to the continuing decrease is uncertain, but the deserted breeding grounds of 6–7 pairs in Glenmore Forest and 3 pairs in Inshriach Forest are now largely afforested (D. Nethersole-Thompson, personal communication). Increasing human disturbance has also played a part, especially around Loch Morlich, the main feeding place for the Glenmore group. Greenshanks still breed in small numbers in the headwaters of the Spey, Findhorn and Dee, and on Rannoch Moor, but their refuge is the wild country north of the Great Glen.

The BLACK-THROATED DIVER is the second bird that I associate especially with the Highland deer forests. Rankin (1947) has described the haunts of our two breeding divers, and emphasised not only that Black-throats prefer the bigger lochs with islands, but also that they occur mainly on those least subject to marked fluctuations in water level. Because of its locomotory difficulties on land, the bird lays its eggs on a low shore, within a few metres of the water's edge, so that it can easily flounder into its more natural element at the approach of danger. Such a nesting site is highly vulnerable to flooding after the prolonged and often heavy rain so frequent in the Highland spring, and in some sites, eggs are quite commonly washed away. An island site is much preferred, as safer from both Man and Foxes, but some pairs will nest on the shores of lochs with none.

Black-throated Divers nest in some of the wildest and most beautiful mountain country in Britain. They are themselves birds of striking beauty, and their far-sounding territorial wail in the spring is entirely in keeping with the wilderness character of their surroundings. Black-throats are usually scattered in single pairs, but some of the larger lochs have at least two pairs. They can nest on quite small lochs (down to 10 ha) (Bundy, 1979) which offer the security of islets, but will then usually range over other lochs for feeding. During the breeding season, this species is, unlike the Red-throated Diver, an inland and fresh water feeder, though it takes almost entirely to the sea during autumn and winter. It feeds by diving, mainly in water no more than a few metres deep, and takes small fish (mainly trout), frogs, crustacea, molluscs, insects, annelids and other invertebrates. The majority of breeding and feeding lochs have nutrient-poor water; although their biological diversity is limited, many are important for their trout and salmon fishing.

During 1900–60, the Black-throated Diver appeared to expand its Scottish breeding range, becoming more widespread in the Hebrides and reaching Arran, Carrick and Galloway (Thom, 1986). During the past 20 years there has, however, been increasing concern about its steady decline in some areas. Sharrock (1976) estimated a total nesting population of

150+ pairs, a figure recently confirmed (Stroud *et al.*, 1990); but RSPB data have shown a consistently low breeding success of only an average 0.29 young per territorial pair, in population samples between 1972 and 1984. In 1987 there was said to be only one pair left in Argyll.

Increased disturbance by fishermen is thought to be one possible cause of this unhealthy situation. Although in earlier times fishermen disliked the bird and often destroyed its eggs for taking their trout, it is not direct persecution which now appears to be the problem. The incubating parent readily takes to the water when humans appear at the loch edge, or in a boat, and if they are there for several hours, the bird may be off the eggs long enough for these to become chilled or predated. Bird watchers may also cause increasing disturbance to some breeding lochs. While this species has always been a bird photographers' favourite, it is a Schedule 1 species for which nest photography has for many years been regulated by a licence and quota system. At this moment, no licences are being issued.

In most of the Black-throated Diver's haunts there is no reason to suppose that predation has increased. The western Highlands have remained the chief refuge of the otter in Britain, but it is unlikely that this animal has increased in inland districts. Encroaching afforestation may bring increased risks of predation (p. 66). In the Boreal regions, Black-throats commonly nest on lakes amidst the coniferous forests, but nest predators such as Hooded Crows are probably here in different balance with their food chain, not least because they are themselves preyed upon, by Goshawks in particular. The evidence for pronounced acidification of lochs in south-west Scotland raises the possibility of adverse effects for this bird, at least in this region, where its nesting seems to have ceased again. Surprisingly, however, Eriksson (1986) found that in the strongly acidified lakes of southern Sweden, Black-throated Divers were apparently unaffected either in numbers or breeding success. The explanation offered was that depletion of the fish stocks had removed an important predator in the pike, and reduced competition for invertebrates on which unfledged young were fed; while in these exceptionally clear waters, food was more easily found.

Whatever the reasons for its decline, this is a species needing all possible protection. One novel approach, by the RSPB and Forestry Commission, has been to build artificial islands in lochs where there were none. These are rafts, anchored to the bed of the loch, but floating so that they are not affected by change in water level. Covered with the vegetation of the natural nesting habitats, they have successfully attracted Black-throated Divers to nest in several localities where breeding did not occur previously. Some purists object to such interference with nature, but when wildlife is under such great pressure from human activities, some modest intervention in the case of declining species seems entirely valid and commendable.

Red-throated Divers are also quite widespread in the deer forest country, especially towards the west and north coast. They are more patchily
distributed and usually on small lochs and lochans, but these can be in
quite rugged mountain situations as well as on level flows. They seldom
appear on a Black-throat loch at the same time as that species. Red-throats
also extended their range southwards during this century, through the
southern Hebrides, to Arran and Galloway. There has long been an Irish
outpost in Donegal, in rugged moorland much resembling that widespread in Scotland. Both Divers have always been scarce or absent over
much of the eastern Highlands, and the Red-throat reaches its highest
breeding densities on moorlands close to the sea, where it mostly feeds.
Some pairs far inland resort to bigger lochs for feeding.

Small numbers of Great Northern Divers had long been known to summer in northern Scottish waters and breeding had been suspected. there
was great excitement when a pair was finally proved to breed in Wester
Ross in 1970, but the event proved to be short-lived (Thom, 1986). This is a
mainly North American diver (the Loon) but nests widely in Iceland, and
appears far more numerous in British and Irish coastal waters during winter than the Black-throated Diver. Its failure to colonise Scotland permanently is surprising, and yet another example of the conservatism in distribution of birds which seem well placed to expand their range.

5

The flows

The tendency to soil waterlogging and peat formation in the cool and humid uplands of Britain and Ireland increases westwards and northwards as summer temperatures fall, and rainfall and atmospheric humidity rise. Plant remains tend to accumulate as peat on level ground once there is an appreciable annual excess of rainfall over evaporation and transpiration. The lowland raised bogs of Dyfed, South Humber, Morecambe Bay, the Solway, the Carse of Stirling, and central Ireland are peat moorlands which formed through the progressive acidification of poorly drained plains often with earlier lakes and fens. On the uplands, peat development is favoured especially on broad plateau watersheds, such as those of Dartmoor and the Pennines. Here, the peat covers all the gentler features in a smothering mantle; hence its apt name, blanket bog. And, as the geographical gradients of climate produce an increasing surplus of water, so the tendency to peat formation steadily overrides slope of ground and descends in elevation. Flat or almost flat areas of both raised and blanket bogs are widely known through Scotland and northern England as 'peat mosses' or 'flows'.

In the extremely wet districts of western Ireland, and western and northern Scotland and its islands, great areas of blanket bog occur down almost to sea level. The most extreme development of this peat-clad terrain occurs in the far north-east of Scotland, where the low undulating moorlands of east Sutherland descend gradually into the plains of Caithness. Here, east of a line from Tongue to Lairg, is one of the largest continuous expanses of blanket bog in the World, covering at least 2500 km^2; for globally, this is an extremely restricted formation, confined to a few cool and oceanic regions. Westwards the peatlands gradually become more dissected, by the steeper and more rugged relief of west Sutherland, with its strange knobbly moorlands and abruptly jutting higher peaks and ranges. The total blanket bog area of Sutherland and Caithness is – or was – around 4000 km^2 (Stroud et al., 1987).

Other important areas of flow ground are in the west of Ireland (especially Co Mayo), on the Isle of Lewis, the Wigtownshire and south Ayrshire moors, the southern Cheviots, and Stainmore in the northern Pennines. Hill plateau blanket bog is widespread and extensive in parts of Wales, much of the Pennines, Cheviots, Southern Uplands, eastern Highlands, Shetland, and on many Irish mountains, often as important habitat within grouse moors (p. 89). These peatlands are, however, amongst the most unproductive and intractable of the uplands as regards land use. Their carrying capacity for sheep is low and though many Highland areas have Red Deer, they are usually in rather small numbers.

The Sutherland–Caithness Flow Country is the classic area of this peatland habitat. Its outstanding importance for birds derives from its large size, northern position and also the variety of open water habitats. Level ground often has extensive development of pools and wet hollows, of varying shapes and sizes, separated by ridges and hummocks of different heights, giving a network of firmer ground. From the air, these pool systems often appear as distinctive patterns, covering up to 1 km^2 or more in any one place (Fig. 51). They are especially well developed in the Flow Country, here covering a large total area, but in numerous distinct patterned bog surfaces scattered over the district. In some places, such as the Knockfin Heights, the average pool size is much larger than usual. The larger pools are called dubh lochans (black tarns) in Gaelic, as befits their appearance of inky depth. Many have a bare peat floor, variably grown

Fig. 51. The Flow Country, near Loch More, Caithness. A 'dubh lochan' system with bogbean, cotton sedge, heather and bogmoss, on the vast blanket bogs of this region. Haunt of Golden Plover, Greenshank, Dunlin, Arctic Skua, Red-throated Diver, Common Scoter and Greylag Goose.

with bogbean and common cotton-grass, but otherwise contain only small algae. Others show colonisation and infilling by aquatic forms of bogmoss. They are the habitat of myriad insects which provide food for some of the peatland birds. Lochans and bigger lochs in a wide range of sizes and shapes are scattered over the Flow Country with varying frequency. The bigger lochs have a sandy or even rocky shore, and many have fringing patches of sedge, representing the beginnings of fen. They often lie in largely peat-covered catchments and belong to the nutrient-poor dystrophic class of lakes.

These moorlands are drained by numerous streams, feeding the main rivers of Naver, Halladale, Shin, Brora, Helmsdale, Berriedale, Thurso and Forss. Originating as meandering peaty runnels amongst the bogs, they course more rapidly over stonier beds lower down. In some of the main tracts of peatland, exposed rock outcrops are quite infrequent, and the only good crags are mainly where some bigger streams have cut gorges. Although the stream waters are mostly poor in nutrients, there are some notable salmon rivers amongst those listed.

The Flow Country is our nearest equivalent to the wet tundras of the Low Arctic beyond the northernmost forest limits. There are similarities of appearance, in the patterned mire systems and the prevalence of cotton-grass and bogmoss. Yet the Arctic tundra is produced by quite different climatic conditions, for over much of this great region rainfall is low, and the bog vegetation flourishes because the moisture is held in summer in the shallow surface layer which thaws above the permanently frozen ground. Britain is far south of any permafrost, and our equally treeless southern tundra results from surface waterlogging in a cool oceanic climate.

The Sutherland–Caithness equivalent has drawn a tundra-type bird assemblage which also represents an outlying southern European occurrence of that found widely on the wet Arctic tundras. From the prevalence of wet ground habitats, waders and waterfowl are especially well represented. Of 17 species of wader regarded as upland (Table 1), only Dotterel and Purple Sandpiper are absent, though recent breeding of Whimbrel is uncertain. Some birds of the drier sheep-walks, grouse moors and deer forests (and their streams), are also commonly present, and those of the maritime hills are also represented. While breeding density is often low, this is, overall, the richest of all our upland habitats in species diversity. And, because of its great extent, the total numbers of some birds represent important fractions of their entire British breeding populations.

Harvie-Brown and Buckley (1887) were the first to mention 'the great flow-lands' of this district, and their importance as the breeding haunt of Red-throated Divers, Wigeon, Scoter, Greylag, Golden Plover, Greenshank, Dunlin and gulls of five species. Yet, for almost another century it remained a virtual *terra incognita* amongst ornithologists. For many years, probably the only visiting naturalists were a few egg collectors in search of the rarer species. The Forsinard Hotel was a favourite base, but even the more vigorous walkers could cover only a limited distance from roads, and many of the remotest parts remained unexplored. Bird photographer George Yeates (1948) wrote vividly and evocatively about the Caithness flows, where he worked on Red-throated Diver, Arctic Skua and Dunlin. Local ornithologists gradually built up knowledge of the area, but more systematic study did not begin until 1979, when the Nature Conservancy Council launched a breeding birds survey of the Flow Country. This survey, later joined by the Royal Society for the Protection of Birds, had to be based on samples of the range of habitats, since total coverage was far beyond the resources available. The findings from 77 plots, using a territory mapping census method, over a size range of 200–1025 ha and covering about a fifth of the total peatland area, are reported by Stroud *et al.* (1987).

The Golden Plover was the most numerous and constant wader, closely followed by the Dunlin. Greenshanks were almost as continuously distributed but mainly at rather low density. Curlew were widespread but rather sparse, compared with many more southerly moorlands; while Snipe and Redshank occurred patchily, but especially where there was flushed, marshy ground or wet grassland rather than typical bog. Lapwings were also mainly towards the edges of the flows, especially on the improved, grassy areas associated with crofting land, either extant or abandoned. The Common Sandpiper was a fairly constant bird of bigger and less peaty lochs and streams, while the Oystercatcher and Ringed Plover occurred with surprising frequency in these habitats, particularly on shingly and sandy margins.

Other waders are elusive rarities, not all nesting regularly, perhaps, and requiring an especially determined search. Maybe they have long been there and have only chanced to be found because more birders have visited the district in recent years. Yet their appearance since around 1950–60 would fit the view that the trend to colder, more backward springs in recent decades has brought an increase in number of northern birds breeding in Britain. Wood Sandpipers appear to nest regularly in small and varying numbers around the Highlands, and a handful of breeding places have been found in the Flow Country. The favourite habitat appears to be a swamp-fringed lochan as the main feeding place, surrounded by flow or drier moorland in which the nest is located, at some distance. The parents evidently soon take their chicks to open water, and are more readily located then, in noisy demonstration. Most breeding haunts are at low levels, but one in the Grampians is at 520 m.

Red-necked Phalaropes are equally rare and perhaps still more sporadic here, but this is not their ideal habitat. The best Phalarope localities in

Scotland and western Ireland are in much more base-rich semi-maritime habitats (pp. 175–6). The dubh lochans of the flows are mostly too acidic, and the bird is evidently limited there to the slightly richer tarns with a greater amount of emergent and fringing vegetation. Temminck's Stint has bred in recent years; a tiny wader easily overlooked. Most recorded British nests have been close to northern lakes or rivers, with typical flow ground evidently less favoured than more lowland and fertile situations (Nethersole-Thompson and Nethersole-Thompson, 1986).

The latest and most exciting of the additions to the breeding bird fauna was the finding of a Ruff's nest with 4 eggs in 1980 by J. Massie (Nethersole-Thompson and Nethersole-Thompson, 1986). This is especially interesting as the first known breeding of Ruff apparently from the northern, boreal forest marsh and tundra population of this mainly Arctic–Subarctic species. The recent recolonisation by Ruff of lowland wet grasslands and saltmarshes in England appears to have been from the much smaller southern population in the Low Countries, where it inhabits simi-

Fig. 52. Wildlife in a northern lochan: Red-necked Phalarope amongst bogbean. The role of the sexes are reversed, with the duller-coloured male taking over incubation and care of the chicks.

lar breeding habitats. In 1974, a male Pectoral Sandpiper displayed over a flow, though this Siberian–North American species has never been known to nest in Europe (Nethersole-Thompson, 1986).

Among the waterfowl, the Red-throated Diver has one of its main British strongholds in the Flow Country. It was present on half of the sample areas, and is likely to number at least 150 pairs (Stroud *et al.*, 1987). This bird is satisfied with quite small dubh lochans, preferring to nest on an island, but otherwise laying on the peaty margin. It accordingly breeds on the patterned bogs as well as the bigger lochs. Black-throated Divers are rare, for they occupy only the larger lochs with an adequate food supply limited to those with islands for nesting. There are perhaps 30 pairs in the Flow Country, representing an estimated one fifth of the total British population.

This district holds one of the original native stocks of the GREYLAG GOOSE, which now has numerous feral populations scattered over Britain and Ireland through widespread release of birds by wildfowlers. Even the Caithness population is now believed to be mixed with semi-domesticated birds returned to the wild, but these feral geese mostly came from eggs taken from the native stock. The Greylag is not a particularly northern species, and was once more widespread through southern Britain as a native bird. It is, however, a distinctive inhabitant of the flows, as the only British breeding representative of the goose tribe which are such a notable element of the Arctic tundra bird fauna. The Greylags nest on islands in lochs and lochans, or in longer vegetation, especially heather, out on a broad flow. They take the goslings to water soon after hatching, and need at least a moderate-sized lochan in their breeding area. A moulting flock of some 1200 Greylag discovered on Loch Loyal in 1986 appeared to represent a substantial part of the Sutherland, and perhaps Caithness, native breeding population (Stroud *et al.*, 1987).

Mallard are widespread on the flows, as they are on moorland generally, but Teal are especially birds of peatlands with dubh lochans, and are the most numerous duck, breeding in drier places in the bogs, not far from water. Wigeon are a widespread northern species, frequenting larger lochans and main lochs. The drakes are conspicuous on the water, but the ducks incubate on the moorland up to at least 400 m from the water's edge, and until they have hatched and taken their ducklings to the loch, they are concealed and difficult to find. Around 80 pairs are believed to nest, about one fifth of the total British breeding population. The most famous duck of the flows is the COMMON SCOTER, a rare breeding species in Britain and Ireland, which has a main stronghold on these wet moorlands. Around 30 pairs are reckoned to nest in average years; about 40% of the British population. Some breed on islands, but others choose sites in longer vegetation mostly well away from the water. They are the more elusive through the males' desertion of their mates once the eggs are laid, to join in small parties on bigger lochs, often far from the nests. The drakes moult early in flocks, and the ducks are entirely responsible for the care of the young. Pintail have bred sporadically and Scaup very rarely on lochs in the district but, though both are important Arctic tundra species, they have not shown a particular association with flow habitats in Britain.

Goosanders are rare, but Red-breasted Mergansers fairly widespread, nesting along the bigger rivers and lochs, often on islands. A notable predator of the Caithness flows is the Arctic Skua, nesting sparingly and usually amongst dubh lochan systems. Some pairs nest in solitary state, but there are small groups of a few pairs or, exceptionally, up to a dozen pairs. Numbers tend to vary, and perhaps move around, and the total is esti-

Fig. 53. The spirit of the Caithness flows: Arctic Skua over dubh lochans.

mated to be now around 60 pairs. A hundred years ago, numbers were evidently larger but the bird then became heavily persecuted by keepers and the larger groups were much reduced. Arctic Skuas have never established more than a foothold in Sutherland (1–3 pairs) (Thom, 1986), where there is a great extent of flow ground looking equally suitable to that in Caithness. The Skua has been supposed to depend partly on its parasitic association with gulls, but the numbers of these nesting nowadays on the flows is quite small. Possibly the bird behaves here as a more typical predator, as on the Arctic tundras, where the species often lives far inland and is independent of the sea (Furness, 1987). The Caithness breeding haunts are, however, within reach of the large seabird colonies between Dunnet Head and Helmsdale, and some of the flow skuas may travel to the coast to harry Kittiwakes, auks and terns (Sharrock, 1976).

Harvie-Brown and Buckley (1887) recorded that the Common, Black-headed, Lesser and Great Black-backed, and Herring Gulls all bred on the Sutherland–Caithness flows. Even then their numbers were being thinned and the colonies broken up by gamekeepers and shepherds. Nowadays, the Common Gull is the most numerous species, nesting in small colonies of up to 20 pairs or so, especially on islands or islets in the lochs and lochans, but also in fringing swamp, or on drier shores, and in patterned bogs. Black-headed Gulls nest here in small groups, scattered widely over

Fig. 54. Common Gull at nest in heather on an island in a Ross-shire loch. The most northern and upland of our breeding gulls. Dennis Green.

the moorlands. They favour islands and treacherous marginal swamp, where the nests have greatest security from ground predators. Lesser Black-backs may still nest sparingly in a few places, but they and the Great Black-backs and Herring Gulls are mainly coastal breeders nowadays in this part of Scotland. These large gulls are now better represented on the flows of Lewis (Stroud *et al.*, 1988).

The Red Grouse is a characteristic flow bird and, when the Grouse Inquiry reported in 1911, several moors in both the Sutherland and Caithness sectors of the Flow Country were mentioned as yielding good bags. Since then, numbers have undergone a general long-term decline, but nevertheless remain at moderate levels in some areas, at least in good years. The Caithness moors have long been regarded as the ideal area for the sport of 'grouse-hawking' with trained Peregrines or Gyr Falcons. The huge expanses of almost flat flow enable the often long chases after quarry to be observed, and distant kills immediately followed up, so that risks of losing valuable falcons are minimised. In hilly country, the strong flying grouse and its pursuer all too often disappear quickly from sight, never to be seen again.

The great flows are, indeed, part of the hunting grounds of wild raptors, in search of grouse and other prey. Both Peregrines and Golden Eagles range widely, the former often cruising at some height, where it is little noticed. Both species have only a few nesting places here, on small hillside outcrops and in river gorges, for good cliffs are scarce. Many pairs, both living on the more rugged hills overlooking the bogs and on the sea cliffs, nevertheless include flow ground within their home ranges. From the importance of patterned bogs for ground nesting Peregrines in Finland, it will be surprising if this bird is not one day found nesting in a dry hummock between pools of the similar northern Scottish bogs. The Finnish bog-nesting falcons are vulnerable to ground predators, but the maze-like character of some of these habitats may give protection, since it would take longer for predators to locate and reach a nest, and the Peregrines would also have more opportunity to attack them. The same would be true of many of the larger patterned bogs in Scotland.

Merlins are widely, but thinly, spread over the district, reflecting the rather low density of their principal prey, the Meadow Pipit. Where they nest in level flow ground, they are also still more elusive than usual. Preferred nesting habitat is still the steeper and drier ground with longer heather, on rocky hillsides or slopes of stream glens. Where available, old Hooded Crow tree nests are sometimes used as nest sites. The total population is probably larger than the 30 pairs estimated at present.

Hen Harriers and Short-eared Owls are also widely scattered over the Flow Country, but it is difficult to count the proved breeders. Both species, especially the Hen Harrier, were often seen on the sample survey areas, but they range widely and territories can be large. Both species tend to hunt the moorland edges, where feeding is better: the rough, rushy areas of improved grassland and old crofting ground where there are more field voles and small passerines. Here, as elsewhere, both benefit from the early stages of afforestation. Kestrels are somewhat similar, for they are in rather small numbers, and often associated mainly with the grassier areas which are most productive for field voles.

Hooded Crows range widely over the moorlands, though the nesting haunts are mostly in the remnants of birchwood and other tree growth in the sheltered glens and along the edges of the main blocks of flow. Breeding density was low over much of the area until recently, since this was one of the most treeless parts of Britain. The new conifer forests will provide an abundance of nesting habitat once the trees are high enough, and will place a larger population of Crows in close proximity to much flow once distant from the nearest breeding haunts. Ravens breed mainly on bigger crags around the edge of the district, or in rocky glens, and forage widely over the flows. Probably some 20 pairs have territories which include peatland.

The Meadow Pipit is the most numerous of the small passerines, acting as usual as main host to the Cuckoo and especially in the areas with abundant heather. Skylarks are surprisingly common over the area, though most abundant on grassier ground. Ring Ouzels, Wheatears and Wrens are thinly scattered, mainly on the steeper, drier and rockier terrain, while Whinchat, Stonechat, Pied Wagtail and Twite occur mostly around the moorland edges, and Dipper and Grey Wagtail are on many of the streams.

Wintering birds are few but have been little studied. The Red Grouse remain and continue to attract predators such as the Golden Eagle, but most of the raptors appear to leave for more productive areas. The Greenland White-fronted Goose resorts to the area during winter, to feed on the succulent leaf bases of cotton-grasses in the softer areas. It is difficult to be sure of numbers over this large area, but up to 400 birds have been counted at once. Winter visitors, such as Whooper Swan and Goldeneye, often linger into spring on the lochs, but are not yet known to nest.

A mere catalogue of the peatland features and birds does little to convey the real character of the Flow Country. It is the most desolate part of Britain; a mysterious and lonely land which could be in some remote corner of the globe. Away in the middle of these almost flat bogs, it is possible to sense the infinite solitude of the vast Siberian tundras, described by the earlier traveller naturalists. The watery mazes of the dubh lochan systems, with their strange yet distinctive patterns, are so similar to features of the Boreal–Arctic bogs and tundras. On clear days distant landmarks of the higher hills rise with striking suddenness from the low peatland plains: Morven and its satellites, the Griams, Klibreck, Loyal and Hope. When the clouds are low, it is a still more forbidding, eerie place in which direction is easily lost and one is glad of the reassuring compass. And when the frequent drenching rain which sustains them drives relentlessly over these great bogs, they become inhospitable beyond description.

The birds bring the flows to life in the spring, but they can still seem dead, empty places in May when many species go silent and inconspicuous during incubation. Mid-June is the time to form the best impression of bird numbers and variety, when most of the young have hatched, and the parents become noisy and demonstrative at human intrusion. Some of the bird calls have a fitting touch of the melancholy: the plaintive piping of Golden Plover and liquid flutings of Greenshank, the weird clamour of Red-throated Divers, and the querulous mewing of Arctic Skuas, which George Yeates described as 'the wildest cry in all nature'.

This is a landscape fascinating to the ornithologist, not least from the challenging thought that such a wilderness offers the chance of making some new discovery. Much of it is still unknown, and who knows what the next year will bring? The remoteness, the weariness of the going on foot, and the frequent awfulness of the weather are unspoken factors behind the information hard-won by the bird surveyors of the flows. The dedication of a handful of people has now given us a fair picture of the variety, abundance and distribution of bird life over this district. It is information of a much-needed and timely sort. For the Flow Country is fast disappearing under a different blanket: of trees.

Small conifer forests were established on an experimental scale in the district many years ago, at Borgie, Strathy, Rumster and around Lairg. Their performance did not lead to any early enthusiasm for expansion, and 30 years ago the prospects for extensive afforestation in Sutherland and Caithness seemed remote. All that has changed. The continuing demand for cheap plantable land, development of machinery for ploughing deep, wet peat, selection of the most suitable provenances of Sitka Spruce and Lodgepole Pine for the unfavourable conditions, aerial application of fertilizers and (when necessary) pesticides, and processing of low-grade timber in particle-board mills, have all combined to focus the attention of forestry interests on these northern moorlands. A great expansion since 1980 has seen 67 000 ha of peatland now afforested or approved for planting.

It is not possible to drain and plant the bigger dubh lochan systems, but ploughing often encroaches so closely to their edge that they may well degenerate eventually. Birds nesting in islands of such habitat isolated within enveloping forest must be very vulnerable to predators. For some species, the wider expanse of moorland in which the dubh lochan systems are set is a necessary part of their habitat. The general effect of afforestation is otherwise as described previously: the bird fauna of the peatlands is entirely replaced after 10–15 years by one adapted to thicket woodland, until clear-felling eventually produces open areas. We cannot predict the exact nature of these, but they will certainly have a much drier and grassier type of vegetation than the original bog.

Much open peatland still remains at present, but the unplanned and haphazard acquisition of planting land according to the vagaries of the land market have already spread afforestation widely over the whole district. There has been heavy penetration of the finest area of all, lying north and south of the railway between Forsinard and Scotscalder. Few major catchments are now without some planting, and the wilderness character of the Flow Country is rapidly disappearing. I count myself fortunate to have seen something of this wild and beautiful country before all the tree farming began, but regret that I did not see more of it. This has become one of conservation's more desperate battles, and the saga of the Flow Country will be taken up again in Chapter 9.

What of the other flow-lands? No others can match Sutherland and Caithness, in either extent or variety and size of their bird populations. Those of western Ireland were the next largest, but woefully short on bird variety, with rather sparse populations of Golden Plover, Dunlin and Red Grouse as the most notable species. Red-throated Divers have a foothold in Donegal, Common Gulls are fairly widespread, and Greenshanks have nested in the west, but the southern latitude and warmer climate appear to discourage northern species. Red-necked Phalaropes and Common Scoters breed in the far west, but in somewhat eutrophic coastal and lowland habitats compared with typical flows. Much of the Isle of Lewis is covered with deep blanket bog and this is the next most important area of flows in Britain (Stroud *et al.*, 1988).

Elsewhere, the main flow areas mostly merge into more typical grouse moor. Golden Plover, Dunlin and Curlew are characteristic waders, but probably the most noteworthy feature is the occurrence of sizeable gull colonies. Black-headed Gulls are usually concentrated on moorland lochs and tarns, especially those with extensive fringing swamps where nests are relatively safe from foxes and humans. Sometimes, colonies spill over onto the surrounding flow: there is a sizeable one on Bowes Moor in the northern Pennines. Smaller colonies based on moorland tarns may be erratic or ephemeral in numbers and occupation. Reasons for desertion are often unclear, but sometimes there are keeper attentions as a possible reason, and afforestation has caused abandonment of some breeding places. A colony of 500+ pairs at the Black Lochs of Kilquhockadale in Wigtownshire disappeared after the surrounding flow was planted. A Common Gull colony on the flow of Leanachan, east of Fort William, also hung on for a few years after planting but then gave up.

Scattered colonies of LESSER BLACK-BACKED GULLS have been a feature of flow ground. There were several on the peat mosses of the

Morecambe Bay hinterland (Foulshaw and Roudsea Mosses), and the Solway Firth (Wedholme Flow, Bowness Moss, Lochar Moss, Anabaglish Moss). Others were on moorlands far inland, such as Butterburn Flow and Hindleysteel Moss in the southern Cheviots, and Gull Nest in Morayshire. Small numbers of Herring Gulls also bred in some of these colonies, and there were often from one to a few pairs of Great Black-backed Gulls as well, with up to 20 pairs on Bowness Moss. Occasional pairs of Great Black-backs have nested in isolation on the Cheviot moors, but keepers have usually put a stop to such attempts.

Nearly all these earlier big-gull colonies have become defunct in recent years, yet another, on the Bowland Fells on the west side of the Pennines above Lancaster, has grown to eclipse all the forerunners. Variously known as the Abbeystead, Mallowdale or Tarnbrook Fell gullery, and believed to originate in 1938, it is scattered over 6 km² of high plateau moorland (400–530 m), mostly on degraded and eroded blanket bog with heather and cotton-grass over Millstone Grit. Some of the colony is on ground with bilberry and grassland, and there is much bare peat and stones in places. Greenhalgh (1974) reported that in 1972, besides 17 000 pairs of Lesser Black-backs and 3500 pairs of Herring Gulls, there were 35 pairs of Great Black-backed Gulls. The colony had grown to an estimated 25 500 pairs in 1979, of which 86% were Lesser Black-backs and 14% were Herring Gulls (Wanless and Langslow, 1983).

This is by far the biggest inland gull colony of any kind in Britain or Ireland. Desultory attempts were made to reduce its size in the interests of Red Grouse management (this is the general area where the record bag was made: see p. 92), but from 1978 onwards more systematic control was initiated to reduce salmonella contamination risks in a water supply reservoir within the catchment. By use of stupefying baits, an estimated 23 400 gulls were killed in 1978, and large numbers of nest contents destroyed, yet the breeding colony was still larger in 1979. Further kills had nevertheless reduced the colony below an estimated 10 000 pairs in 1982. Density of nests had previously varied from 0.63 to 0.15/100 m² over different parts of the colony and an effect of concentrating control effort in areas of high density was to even out nesting density to a lower average figure of 0.16 nests/100 m² (Wanless and Langslow, 1983). Continued culling had reduced the breeding population to just over 4000 pairs in 1988, but the colony continued to occupy the same 6 km² of moorland (Thomas and Tasker, 1988).

There seems to be no particular reason why this part of the Bowland Fells should have proved so attractive to these big gulls; perhaps a decline in keepering during 1940–45 may have been the chance event which allowed the original colony to become firmly established. The location, on a little-visited tract of high moorland within 20 km of the sea, may also have given certain advantages; and the colony developed during a period when coastal populations of these gulls were increasing generally. One interesting ecological effect of large inland gull colonies is the enrichment of otherwise poor ground by a fairly substantial transfer of nutrients from excreta and food remains. Nitrogen and phosphorus levels are enhanced, and the numerous discarded shells of crustaceans and molluscs add calcium. On parts of the Abbeystead colony's ground there has been a notable increase in grasses as a result. Eutrophication by the large (4000 + pairs) Black-headed Gull colony at Sunbiggin Tarn on Shap Fells is seriously reducing the botanical interest of the vegetation bordering this limestone tarn. Probably in the majority of all upland gull colonies, the bulk of the feeding is done away from the nesting area, or even off the moor altogether. The moorlands and their lakes are evidently used mainly as relatively secure nesting habitat.

Finally, the largest colony of Common Gulls in Britain (4000–5000 pairs) occurs on high level flow ground of the otherwise dry and heathery Correen Hills in Aberdeenshire, in company with 80 pairs of Lesser Black-backed Gulls, 40 pairs of Herring Gulls, 3 pairs of Great Black-backed Gulls and several pairs of Common Terns (Nethersole-Thompson and Watson, 1981). Another group of at least 1000 pairs occurs on moorland in the same district, these unusually large concentrations being attributed to the proximity of good feeding grounds as well as undisturbed habitat (Thom, 1986).

6

The maritime hills

In the far northern island groups of Orkney and Shetland, and in the Hebrides, are low hills and moorland strongly influenced by the sea. They and their birds are northern and upland in character, but have some distinctive coastal features and species. Their climate is strongly oceanic, giving much damp, misty weather and peaty soil, and their extreme windiness is reflected in a scarcity of woodland, both natural and planted. Summers are cool but snow cover seldom prolonged. Montane dwarf shrub heaths occur down to 180 m, and the summit of Ronas Hill in Shetland at 450 m has a granite fell-field even more barren than that of the high Cairngorms. Much of the vegetation is moist grassland or heath, but some gentler ground has an extensive covering of blanket bog.

On these windswept islands, salt spray is carried some distance inland, and on exposed headlands it produces a vegetation akin to saltmarsh swards. The sea has an ever-present ambience. Many moorlands end abruptly in high sea cliffs, often crowded with clamouring hordes of breeding seafowl: Puffins, Guillemots, Razorbills, Kittiwakes, Herring Gulls, Fulmars and, more locally, Gannets. The babble of sound and reek of guano waft up from below as one follows the edge, and the air in front of the precipice is thick with birds. In places, colonies of Common, Lesser and Great Black-backed Gulls occupy part of the moorland. Some species breeding wholly on the moorland also have a close dependence on the sea or its birds, and give a distinctive faunal character to these maritime hills.

Orkney is outstanding for moorland birds. Its prevailing Old Red Sandstone gives relatively fertile soils, and a good deal of quite productive farmland, with cattle pastures and even some arable. There was formerly also a great deal of rough, marginal grassland and heath around the edges of the higher unenclosed moorland and bog. Orkney is famous for its ground-nesting birds of prey: Hen Harrier, Short-eared Owl, Merlin, Kestrel and Hooded Crow. The last two species are evidently encouraged in this habit by the absence of mammal predators.

The Orkney moorlands became a crucial last stronghold of the Hen Harrier during the earlier part of this century, when game preserver repression came close to eradicating the bird from the British mainland. Through the dedicated protection efforts of a few Orkney enthusiasts, notably George Arthur, Duncan Robertson and Eddie Balfour, the population here steadily increased from the 1930s onwards. Numbers rose to 1950, fluctuated during the 1950s and 1960s, and peaked in the mid-1970s, but have since declined again (see p. 123 and Cadbury 1987). Polygyny has always been a feature of the Orkney population, so that it is inappropriate to speak of a total number of breeding pairs here.

The high numbers and breeding density of Hen Harriers on Orkney suggest an unusually good food supply. Donald Watson has pointed out how the good numbers of the species on Mainland Orkney, the Uists and Benbecula and their virtual absence from Shetland, Harris and Lewis, correlate with the presence and absence of voles on these two groupings of islands. The same is also true of the Short-eared Owl, which fluctuates in numbers. The large amount of marginal land improvement and moorland reclamation for agriculture on Orkney during the last decade is evidently a problem for these predators which hunt both types of habitat. This land use change has evidently reduced carrying capacity for those upland species which depend a good deal on the ground beyond the moorland edge for their feeding. As well as Hen Harriers, Merlins have decreased, from about 25 pairs in the 1950s to only 5–10 pairs in 1985 (Bibby and Nattrass, 1986). Golden Plover and Dunlin have also declined (Booth, Cuthbert and Reynolds, 1984) and the especially large Dunlin population which once bred around Lochs Stenness and Harray is much reduced. The once good stocks of Red Grouse on some Orkney moors have decreased considerably.

The Shetland group is composed mainly of acidic schists and gneisses, which give a prevalence of infertile and peaty soils, and there is rather little farmland. The island of Yell is largely covered by blanket bog and wet heath. Although carrying capacity is generally rather low, these are interesting and productive moorlands for birds, and again owe much to the influence of the sea. The northern islands of Unst and Fetlar also have areas of serpentine which produce more base-rich soils. Red Grouse are very localised, and absent from the northern isles, while Golden Plover and Dunlin appear to have declined over the last century. The moorland raptors are poorly represented on Shetland, with Kestrel, Hen Harrier and Short-eared Owl absent as nesting birds. The Merlin is widespread though at rather low density, and has declined from 25–30 pairs in the 1970s to 20 pairs in 1985 (Bibby and Nattrass, 1986). Yet, while the Great and Arctic Skuas are now widespread nesters on the Orkney moorlands, Shetland was their great nineteenth century refuge from which they later spread out to other areas.

The GREAT SKUA is one of our most interesting birds, as a globally restricted species which belongs mainly to the fringe of Antarctica, the islands of the southern oceans, and south-east Argentina. It is everywhere mainly an island bird breeding on treeless oceanic tundras, and showing a general dependence on the sea. In the northern hemisphere it breeds mainly in Iceland, the Faeroes and northern Scotland, but has spread north-eastwards recently, reaching Svalbard. The northern population is evidently an offshoot of the southern one, derived from relatively recent colonisation by birds which wandered north. It has not been isolated long enough to have diverged in any notable way from the southern form, and although its Shetland name of Bonxie is said to be of Norse origin, Furness (1987) believes there is no evidence for its occurrence in Scotland before the mid-eighteenth century. Great Skuas nest mostly in rather loose colonies or groups, but on the island of Foula there is a great colony, numbering 3100 pairs in 1977. This powerful seabird is pirate and predator, but also a fish-eater in its own right. It parasitises the colonies of cliff-breeding seafowl, robbing them by chasing until they drop or disgorge their food. The bigger gulls are not exempt from this treatment, while Kittiwakes and birds of similar size are also taken as actual prey. Great Skuas are also great robbers of young and eggs of most birds around them, including their own relative, the Arctic Skua.

Because of its predatory habits, the Great Skua has never been popular with sheep farmers, though probably most of the mutton in its diet comes from carrion. It was a persecuted species, to the point where decrease and retreat limited its distribution to Foula and the northern tip of Unst at Hermaness. By 1831, little more than 100 pairs remained in these two groups, most of them on Foula. Here and in the Faeroes, both the eggs and young provided human food. Then, in later Victorian times, eggshells and skins were at a premium amongst collectors. Protection of the Hermaness colony by the Edmonstons, who owned the land, led to a slow but steady increase, and the species has gradually spread out through most of the Shetland and Orkney Isles, and reached as far west as St Kilda. Scattered footholds have been established along the Caithness and Sutherland coast; breeding is sporadic here, though pairs are now prospecting inland flows. The population was still increasing in 1987, when it numbered around 7900 pairs (Stroud *et al.*, 1990); its healthy status represents another success for bird protection.

Great Skuas nest mostly on wet grasslands, heaths and shallow blanket bogs of these oceanic islands and coastal headlands. Human intrusion into a nesting area is marked by low, diving swoops by the Skuas, which become increasingly menacing as a nest is approached. The parent birds come hurtling at the head of an intruder, skimming past with a truly disconcerting rush of sound (Fig. 55). Some especially aggressive individuals will repeatedly press home their attacks, to strike people with heavy blows, if given a chance. As soon as one territory is left, the owners of the next will take up the onslaught. Where the nests are close together, a person may come under multiple attack, and it takes nerves of steel not to duck. When not persecuted by man, this appears to be a successful species, in that it has a small clutch size (usually 1–2 eggs against the 3 of most gulls), does not breed until 4–6 years old, and usually contains a substantial non-breeding element in the population. This suggests a low adult mortality, a characteristic of other seabirds which spend their lives outside the breeding season on the open ocean, as does the Great Skua.

The ARCTIC SKUA is often an associated species on these maritime moorlands. While this is a notable bird of the Caithness flows, most of its British population nests in Orkney and Shetland, where there are close to 3000 pairs. Scattered smaller colonies occur in the Outer Hebrides, Coll and Jura. This smaller Skua has also increased after a Victorian period of numbers depressed by persecution. It has expanded rather more widely than the Great Skua, but not reached so large a population. Although the two often share the same nesting habitats, its larger relative may reduce the Arctic Skua to some extent, if only locally. This happens where the nesting habitat is limited, as on the islands of Noss and Foula, but on larger areas of moorland, the Arctic Skua has less problem in moving to other

Fig. 55. An unwelcome intruder: Great Skua in aggressive mood.

ground when displaced.

This is one of three widespread circumpolar skuas of the Arctic tundras, but is much less associated with lemmings than the Pomarine and Long-tailed, which do not breed in Britain. It is parasitic, robbing seabirds of their food, especially Arctic Terns and Kittiwakes, but even species as large as the Gannet; it is also a predator, taking small passerines and other birds such as waders up to the size of Golden Plover, and small ground mammals. Eggs of various species are freely taken, and carrion when available. With the addition, at times, of insects and berries to its diet, this bird has some claim to be termed omnivorous.

The Arctic Skua is another aggressive species in its nesting haunts, as befits its general demeanour. Other intruding predators such as eagles and falcons are chivvied from the territory, and humans are subject to the same kind of intimidatory 'dive-bombing' as that practised by Bonxies. Sheep and lambs are also hustled from the vicinity of the nests, and this gives an additional cause for annoyance amongst shepherds, as on Fair Isle, where demands for a reduction in Arctic Skua numbers have resulted. This skua is also given to an energetic injury-feigning distraction display at humans when there are young nearby. The range of nesting habitat is rather wider than for the Great Skua, ranging from wet, pool-studded *Sphagnum* flow (Fig. 53) to dry, short heather ground, but including the same shallow blanket bog and moist grassland favoured by the larger species.

This skua has the additional interest of occurring in two colour phases, a uniform dark type, and a light form with pale face, nape, neck and underparts, with a series of intermediates. In Britain dark birds and intermediates are the prevalent forms.

The Shetland bird par excellence is the WHIMBREL, for 95% of the whole estimated British population of about 500 pairs nests here (Richardson, 1990). Only a handful of pairs breed elsewhere, in Orkney, the Outer Hebrides and the northern Scottish mainland. This smaller, Arctic-Boreal version of the Curlew appears to have a southern climatic limit which only just touches the northern tip of Britain. It may be temperature-limited, and Sharrock (1976) noted that the Whimbrel declined in Shetland and disappeared from Orkney during the period of warmer climate from around 1900 to 1950; but increased at least threefold since the trend to colder, backward springs set in around 1960. The continuing increase, now 8.5 fold and at 7.5% annually, is also attributed by Richardson largely to the cooling of climate which lasted up to at least 1980. He believes that internal recruitment must have been boosted by immigration from outside Scotland to account for the increase. The Whimbrel's absence from higher-level moorland on the Highland mainland suggests that a day-length adaptation might also be involved in its restriction largely to the far north of the region.

The Whimbrel's habitat does not seem particularly special or limited. Some pairs nest on the rather distinctive maritime and serpentine heaths in which short heather is mixed with grasses and small herbs, and sometimes abundant moss. Many pairs are, by contrast, on shallow blanket bog with cotton-grasses and *Sphagnum*, sometimes on the ground broken by pools or erosion channels, or even among old peat cuttings. Both these habitats, but especially the second, are widespread on the Scottish main-

Fig. 56. Breeding distribution of Whimbrel in Britain and Ireland during 1968–72 (from Sharrock, 1976).

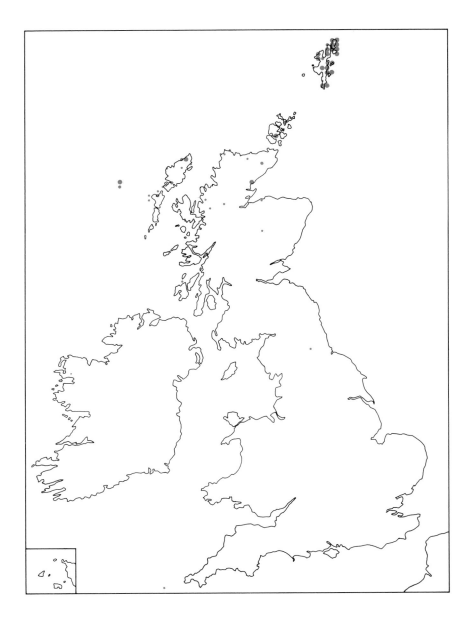

land, and so the bird's absence there is not attributable to lack of suitable terrain. Richardson (1990) found that, in Shetland, short-sward heath and grassland was the preferred habitat, with densities of up to 17 pairs/km^2, contrasting with 6 pairs/km^2 on blanket bog. He also noted a tendency for Whimbrel to group into loose aggregations of more than 10 pairs, with solitary nesting by only 8% of the population. Descriptions of Icelandic and Scandinavian nesting grounds also show that Whimbrel nest on a wide range of sub-montane heaths, grasslands and bogs (including forest bogs), many of which have Highland equivalents. The food of Whimbrel in Scottish breeding haunts has not been closely studied, but is said in Scandinavia to include a good deal of vegetable matter, including moorland berries, besides invertebrates.

Although Whimbrels and Curlews sometimes breed in close proximity in Shetland, the latter tend to occupy the damp pastures and rush fields of the lower slopes and valley bottoms (Williamson, 1951). Their nesting is somewhat separated in time, for the Whimbrel lays about a month later than the Curlew, but they appear to have rather similar food preferences. The two species may thus compete to some extent, in which case the larger Curlew might be supposed to have the advantage. The Whimbrel is an aggressive creature with other trespassing birds within its nesting haunts, however, ousting Crows, Ravens and even Arctic Skuas in determined attacks, often made by the pair in unison. It is often bold with humans when there are unfledged young, demonstrating and scolding within a few metres distance. In this it shows a trait developed by several far northern waders, of tameness to Man. Whether this reflects the general remoteness and lack of human presence in the main breeding grounds is not clear, but it seems an improbable characteristic in the face of appreciable persecution.

The RED-NECKED PHALAROPE is another charming and confiding Arctic-Boreal wader of the northern Scottish islands (Fig. 52). It is another fringe species but a rare and declining one, with only an estimated 19 pairs in 1988 (Stroud *et al.*, 1990). There was formerly a stronghold on the machair lochs and marshes of Tiree, South Uist, Benbecula and North Uist, but numbers are now at a low ebb there. The remaining refuge is on Fetlar in Shetland. There was in 1900 a strong colony on the Mullet peninsula in the far west of Ireland, also in wet habitats among coastal dunes, but this is reduced to a few pairs or may even be extinct.

This is a bird once regarded as especially threatened by egg collectors, and at risk of extinction in Britain through their activities. In recent years it has received good protection in most breeding haunts, yet has continued to decline slowly. Some of the machair habitats have deteriorated through draining which has lowered loch levels and dried out some of the marshes. One Caithness flow breeding place has been deserted through afforestation. Yet the reasons for decline elsewhere and for failure to colonise other apparently suitable habitats are unclear.

While the Red-necked Phalarope breeds widely in Scandinavia, Sharrock (1976) points out that this population migrates south-eastwards, while those of Iceland and Greenland go to the east coast of North America. The British population may thus receive little external replenishment, and perhaps depends on its own performance for maintenance of numbers. An adverse balance between mortality and recruitment may have developed. The species does not seem to have responded favourably to the recent trend towards colder springs, in the same way as the Whimbrel and some other northern birds. This is perhaps not surprising, if the temperature effect works mainly by latitudinal deflection of the migration stream. It is one of our latest summer visitors to arrive and the Venables found that, with hatching dates from 21 June to 5 July, the Shetland Phalaropes were no earlier than those in Greenland (Bannerman, 1961).

The Red-necked Phalarope breeds high in mountain marshes of the upper zone of low willow scrub tundra in Norway, but it is always a low-level bird with us, and its main Scottish and Irish haunts are close to the sea. Few nests have been found above 100 m elevation, so that the preference is for habitats akin to low level northern tundra. Most British and Irish breeding localities have open water feeding habitats which are markedly richer in dissolved nutrients than pools and lochans on acidic moorland. Some inland habitats are akin to 'poor fen' with oligotrophic waters only slightly richer than those of typical blanket bog pools, but these usually have only a pair or two of Phalaropes. The blown shell sand marshes and

Fig. 57. Red-throated Diver on a Scottish hill lochan. These aquatic birds often commute to the sea to feed. Robert T. Smith.

lochs are highly calcareous, and those on the Shetland moors are evidently enriched from sea spray and base-rich soils within the catchment. The more eutrophic conditions are presumed to give a greater biomass of invertebrate food, though certain preferred groups may be especially favoured. The scarcity of these good habitats within its geographical and climatic breeding range must be a basic limitation for this bird in our islands, increasing its vulnerability to other adversities.

The RED-THROATED DIVER is one of the most notable birds of the Shetland moorlands and probably reaches a higher breeding density here than anywhere else in Britain. Bundy (1976) found 39, and possibly 41, pairs nesting in Unst in 1974: as about 80 other possible nesting places had no divers, this population may have been close to saturation, though numbers have evidently increased markedly since nineteenth century persecution abated. Bundy noted that the birds preferred to nest by lochans and pools of less than 1 ha, 96 out of 120 nests being so situated. On Foula, where there is also a high density, the adults have been proved to be highly faithful to their nesting locations in successive years (Furness, 1983). On the Scottish mainland and the Hebridean islands with Foxes, Red-throated Divers prefer lochans with islets on which to nest, but lay their eggs on the shore where there are none. The rudimentary nest is usually within a metre of the water.

The nesting tarns are usually devoid of fish, and their invertebrates are insufficient food, so that most breeding divers fly to the sea to feed. One of the unforgettable sounds of Red-throat country is the weird clamour of the commuting birds as they fly high overhead. On the maritime hills, such as those of Shetland and Orkney, all the nesting pairs are within easy reach of the sea, so that their feeding habits have a relatively low energy cost. On the other hand, many pairs have here to contend with a fearsome concentration of potential predators, in the form of colonies of all three large gulls and both skuas breeding in the immediate vicinity. Some of the diver nesting places are right in amongst the territory of these dangerous neighbours.

In 1974, Bundy found that 22 of 39 first clutches were lost, plus another two empty nests where eggs were never seen. There were 14 repeat layings, but only 16 pairs eventually fledged young: a total of only 18, or 0.46 young per pair. There was little direct evidence of predation and, although circumstances pointed to the skuas and gulls, Hooded Crows were also suspected of stealing eggs and Otters of taking young. The wonder is that, given the seeming level of hazard, any young are reared at all in such areas. The reality must belie appearances. Divers' bills deserve respect, and the adults may normally be able to ward off attacks on their nests or young. There was some indication that casual human disturbance by fishermen and peat-cutters, and also the intrusions of bird watchers, might make the eggs or even young more vulnerable to predators by moving the parents away. Earlier nests were noticeably the more successful. The adults probably have a low mortality, though Bundy found the eaten remains of three possibly killed by Great Skuas, and they have been known to fall prey to Peregrines.

As soon as the young are strong enough to fly, the whole diver family takes to the sea, and the moorlands are deserted until the following spring. The bird then winters in coastal waters between Scotland, Ireland and France, mixing with the more migratory north European population.

The coastal fringes of the hill country in Scotland and Ireland now appear to be the main habitat of the Twite. Many pairs nest in high scarps and rocky gullies overlooking the sea, but others utilise a variety of habitats in the coastal crofting lands, and some build in the favourite Linnet sites of gorse bushes. In the far north and west the Twite is as much a lowland as an upland bird, yet even here it is declining locally (Thom, 1986). For such an apparently adaptable species, its status and distribution are a puzzle (see pp. 115–116).

Other upland species of dry, rocky ground, such as the Merlin and Ring Ouzel, nest in places along the coastal slopes and cliffs abutting the northern and western uplands. The wide breeding distribution of birds such as Peregrine, Kestrel, Buzzard and Raven on seacliffs far from hill country shows that these are as much coastal lowland as upland species. The Chough is also a bird of southern, maritime heaths and grasslands rather than true mountain country in Britain and Ireland. The Sea Eagle formerly had some coastal cliff haunts which were away from the hills, but the Golden Eagle appears with us to be a true upland bird, for its seacliff breeding places are virtually all where the northern moorlands reach down to the coast.

The last upland bird to be included for maritime hills is a surprise, and an almost fortuitous connection with the coast, since its main habitat is the inland tundras and fell-fields of the Arctic. The finding by Bobby Tulloch of a pair of SNOWY OWLS nesting on Fetlar in the Shetlands in 1967, was one of the most exciting events in the recent history of British ornithology. The large clutch of 7 eggs produced 5 flying young: the first ever known to be reared in this country. Valerie Thom (1986) has summarised the subsequent history of the birds in this locality. During 1967–75 successful breeding took place in eight different years and 20 young were reared, apparently by the same pair. An imbalance in the sexes developed, beginning with two females paired to a single male from 1973 to 1975, but with desertion of the second clutch each year, since the male could not supply enough food to both mates. By 1976 the male had gone and since then from two to four females have been present; sometimes infertile eggs have been laid, but another male has yet to appear. The imbalance extended to the young, with numbered 14 females and 6 males.

The numerous recent records of Snowy Owls around the Shetlands are likely to be mainly of these progeny, but since 1970 there have also been sightings in Fair Isle, Orkney, St Kilda and the Outer Hebrides. The trend to colder springs since 1960 has evidently deflected an increased number of birds southwards from their main Subarctic and Arctic breeding grounds, but even on our northernmost islands and mainland hills conditions are evidently marginal for the species. It appears to be a good example of a bird striving to achieve a new foothold towards extending its range, but hardly managing to sustain this, because of limiting conditions. The Fetlar nesting habitat is on rocky sub-montane grass-heath, a terrain widespread in northern Scotland, and there is probably no shortage of tundra-like habitat which would suit Snowy Owls. Food is perhaps a more serious problem. There are no lemmings, but the Fetlar owls quickly adapted to a diet largely of Rabbits, and then switched to seabirds when myxomatosis removed this prey. There are no Ptarmigan (the other favourite Arctic prey) in the island groups where Snowy Owls mostly appear, but they are probably the attraction for the occasional owls which summer on the high Grampians and Cairngorms. It may be that conditions are not quite right in all the necessary aspects, but if sufficient birds of both sexes are still failing to reach our shores to maintain pairs of breeding age, this would explain the precariousness of the present foothold.

7

The high tops

The last of the mountain habitats, the high ground above the climatic limits to tree growth, has a special fascination. It is on the high tops that conditions approach most closely to the severity of the Arctic regions beyond the tree-line, or to the snowy mountains of continental Europe. Here, we may have just a glimpse of those inhospitable lands where life is close to its limit, before the polar deserts of permanent snow and ice finally take over. The stony fell-fields and snow-bed hollows of the Highland summits are fragmentary outliers in our oceanic and temperate region of the vast areas of such terrain in Svalbard, Greenland, Novaya Zemlya, the Taimyr Peninsula, and the islands of the Canadian Arctic. They also occur extensively much closer to Britain, in the mountains of south-west Norway, beginning with the great plateau-land of the Hardangervidda.

The post-glacial restriction of this montane or arctic–alpine zone to the high tops, and its tendency to occur at increasingly low elevations in a north-westerly direction, have been described in Chapter 1. By far its greatest extent is on the high tableland of the Cairngorms, which has much the largest area of ground above 900 m in Britain. These montane habitats vary across the uplands and perhaps the most typical examples occur on the rather lower summits of the central Grampians around the Pass of Drumochter. These are heathery hills, with high-level grouse moor on their flanks. At around 750 m on wind-exposed spurs and summits the heather grows as a prostrate mat, often with abundant lichens and other dwarf shrubs such as least willow, northern crowberry, bearberry and dwarf azalea. On the steeper and more sheltered slopes, the first sign of longer than average snow cover is the replacement of heather by dense patches of bilberry (blaeberry). Still more prolonged snow is marked by dominance of mat-grass, and the latest snow hollows of all are usually grown with a carpet of moss, liverwort and lichen, with sparse small grasses and herbs.

On the exposed high ground above 750 m, the dwarf heather mat is increasingly replaced by carpets of woolly fringe moss (*Racomitrium lanuginosum*). This moss has a very wide World distribution, but is particularly favoured by the cool, oceanic British climate. On some of these Grampian tops, and on Ben Wyvis in Easter Ross, it covers whole square kilometres of high plateau, as a thick, yielding carpet, either quite level or thrown into innumerable low hummocks. The mountain sedge often grows thickly through the fringe moss carpets and there is usually a sparse growth of small shoots of bilberry, cowberry, viviparous fescue, and bent. As shelter increases and snow lies deeper, mat-grass appears abundantly. Farther east in the Grampians, lichens of the reindeer moss group appear in increasing abundance, and sometimes largely replace the fringe moss carpets by a lichen heath. On the loose, gravelly granite soils of the Cairngorms, fringe moss is abundant but seldom as large continuous areas, and above 1000 m the most characteristic vegetation is an open, tussocky growth of the grass-like three-pointed rush.

On the high tops of the north-west Highlands the *Racomitrium* heaths typically contain an abundance of one or more herbaceous plants forming dense flat patches or cushions amongst the moss: thrift, moss campion and mossy cyphel. Many of the northern and western Highland summits are rocky and the moss heaths are less continuous than on the massively rounded Grampian tops. Instability from both wind-blasting and frost–thaw soil movements also produces more bare ground, sometimes with patterning in the form of terraced slopes, stone nets and stripes, and bare ground of irregular form.

These inhospitable heights are the least productive of the mountain habitats for variety and numbers of birds, but they have several species of outstanding interest. The massive, flat or rounded tops and upper spurs are the best for most species, rather than the more dramatic sharp peaks and ridges, which may have only a small area of montane habitat. Some of the characteristic birds from the lower hills are well represented here. On the bigger plateaux there are nearly always Golden Plover, usually rather sparingly distributed and nesting later than on the lower moorlands. The terrain approximates to that of fell-field and montane heath habitats occupied by predominantly dark-fronted birds in the Norwegian mountains. These high-altitude pairs seem to depend on the upper levels of the mountains for all their needs, resorting to richer feeding places on flushed and wet ground here, instead of descending to feed on the upper fields. A few pairs breed as high as 1070 m. Nest sites are often in fringe moss or lichen heath, or where mountain sedge and short grasses grow abundantly. Where shallow and often eroded bogs, with *Sphagnum*, cloudberry, crowberry and bog whortleberry occur on the high watersheds, they are much favoured by Golden Plover also. Such ground is extensive on the elevated plateaux of the Monadhliath, the Forest of Atholl and the Clova Hills in the Grampians.

The high mountain plateaux are also a favourite haunt of Dunlin, which are patchy in frequency and occur usually as scattered pairs or occasionally as loose groups. There are none on the high tops of the Welsh hills or Lakeland, but Dunlin nest in moderate numbers along the 600+ m plateaux of the Pennines from Craven northwards. Several pairs used to breed on the peaty summit cap of the Cheviot, but there are very few on the higher ground of the Southern Uplands. In the Highlands, the massive tablelands of the Grampians and Cairngorms are their main strongholds as high-level breeders. Dunlin on the high tops are less restricted to spongy bogs than is usually the case on lower moorland. They nest on areas of moss or lichen heath, but usually where the ground is grassy or has abundant sedge; among sparse cotton-grass which indicates the beginnings of peat formation; and on deeper high-level blanket bogs.

The Red Grouse goes quite high where heather ground is extensive and reaches its natural upper limits, some 150–200 m above the potential tree-line, as it does in many parts of the Highlands. Numbers thin out as the heather growth becomes shorter within this montane zone, yet this was probably an important part of the species' natural habitat, when forest once covered so much of the lower hill slopes. In the Pennines, small numbers nest on grassy plateaux above 600 m where crowberry and bilberry are the main sources of food. Meadow Pipits breed commonly up to at least 1000 m, though their numbers decrease at higher elevations. On grassier hills Skylarks also go high, and their nests are typically placed among mat-grass tussocks up to 900 m. Wheatears breed in bleak screes and in crannies beneath or among blocks well into the montane zone at over 1000 m. There are also often Ring Ouzels high in the rocky corries: the males pipe their triple note from many an elevated crag, and I have seen a nest at 900 m on the Caenlochan cliffs. In one place or another, many of the birds which breed mainly within the submontane zone are to be found nesting much higher, and the upper limits are given in Table 1. The Golden Eagle, Peregrine, Raven, Snipe, Common Sandpiper and Wren all nest in small numbers at over 750 m.

Only three species long established as breeders are confined to the montane zone of our hills. The PTARMIGAN has special interest, since it is resident and thus a true survivor from the period, some 10 000 years and more ago, when the Ice Age still held this island in its grip. Along with the many plants which gradually retreated to the high tops and have survived there, the Ptarmigan has hung on, and even flourished, within the now greatly restricted areas of suitable habitat. In common with most of our relict montane plants, too, it is a true Arctic–Alpine species, occurring widely in the high ranges of central Europe as well as through the Holarctic Region. Compared with the Red Grouse, the Ptarmigan in Britain has shown less evolutionary divergence and remains substantially the same bird as that elsewhere in Europe, in both plumage and ecology.

Some taxonomists have recognised a Scottish race, *millaisii*, which differs from more northern forms in having an intermediate, grey autumn plumage and less complete development of the white winter phase (Bannerman, 1963).

The Ptarmigan depends on dwarf shrubs for food, and its disappearance from mountains south of the Highlands is attributable to the general eradication of high-level dwarf shrub heath through heavy sheep grazing and indiscriminate moor burning. There is evidence of its occurrence on the Lakeland fells until the late eighteenth century and on the Galloway hills until 1826, but Ptarmigan have long gone from these districts. Attempts at reintroduction have been unsuccessful, since these mainly grassy hills are now unsuitable habitat. The bird occurs in most of the high ranges of the Highlands, but becomes more uncertain in occurrence on isolated outlying hills. It has not been seen recently on the Ben Griams in east Sutherland, but was recorded as breeding on Ben Rinnes in Aberdeenshire and Morven in Caithness during 1968–72. In the islands it now occurs only on Skye, Mull and Arran, and has disappeared from Rhum and Harris. There is no certain history of the species in Ireland at all. The Ptarmigan is a fairly sedentary bird, but capable of local movements between hill massifs, and has recently recolonised Arran (Thom, 1986).

The Ptarmigan has some claim to be the hardiest bird on Earth. It occurs as far north as 83° 24′ N in Greenland, and endures the appalling severity of winter throughout the high Arctic region, with *mean* January tempera-

Fig. 58. Incubating Ptarmigan in a high corrie of the Wester Ross hills. The tight sitting female blends superbly well with the mountain tundra of sparse vegetation and lichen-grown rocks. Derek Ratcliffe.

ture as low as −40°C. Even in these inhospitable wastes the Ptarmigan appears to move little in seeking to escape the terrible cold, but burrows into the snow for shelter and access to its buried food supply. Scottish haunts must be mild by comparison, for at 1200 m, mean January temperature is only −5.5°C; even though the extreme windiness of the Highland climate gives additional adversity in winter, the Ptarmigan here would seem to have a relatively easy time. Nevertheless, except on ground which is blown clear, snow-lie within the Ptarmigan zone in the Highlands is often continuous from at least mid-December to mid-March, and may extend to four months or more in snowy winters.

Adam Watson has made a close study of the Scottish Ptarmigan and I have drawn on his research findings (Watson, 1963, 1965; Nethersole-Thompson and Watson, 1981). Ptarmigan habitat is the zone of montane dwarf shrub heath and, above this, the high summit heaths and fell-fields with fringe moss, lichens and three-pointed rush. The most favoured food is the leaves and shoots of crowberry, bilberry, heather and least willow, and the fruits of berry-bearing plants. The bird frequents high corries, steep upper slopes and windswept spurs and plateaux, but reaches greatest numbers on rocky ground with block screes and boulder fields which provide abundance of both song-posts and shelter from weather, predators and aggressive neighbours. Nesting occurs in all these situations, and while many Ptarmigan lay their eggs in open, windswept

Fig. 59. Suilven and Inverpolly Forest, Wester Ross & Sutherland. The rugged deer forest terrain of the north-west Highlands, with patchy birchwood, numerous lochs, knobby moorlands of Lewisian gneiss and higher peaks of Torridon sandstone and quartzite. The haunt of Golden Eagle, Greenshank and Black-throated Diver.

places, they are more prone than grouse to choose nest sites in steep, rocky situations, even on the ledges of high crags.

In the Cairngorms and central Grampians, Ptarmigan mainly breed above 760 m but occasionally as low as 600 m. There is a slight overlap with the Red Grouse, which exceptionally nests up to 1040 m in sheltered places, but is mainly below 825 m. The highest recorded Ptarmigan nest was at 1265 m on Ben Macdhui. Ptarmigan quite faithfully follow the altitudinal descent of the montane vegetation zone in a north-westerly direction within the Highlands. Breeding is general down to 550 m on the hills of Wester Ross, while on the low, windswept hills of the Parphe in the extreme north-west corner of Sutherland, Ptarmigan nest as low as 300 m and are seen down to 180 m. Their most characteristic habitat, the prostrate montane dwarf shrub heath, is widespread at these low elevations in this area.

The Ptarmigan shows the same general tendency to fluctuations in numbers as the Red Grouse, but with roughly a decade to each cycle. From a sample study area of 500 ha, Watson (1965) estimated that, over the whole 256 km² of suitable montane habitat in the main Cairngorms massif, breeding population fluctuated between 1300 birds in a low year to 5000 in a high year. Including young of the year, numbers were around 13 000 in a peak autumn. In the exceptionally good years 1971–72, these figures were almost doubled. These peak densities far exceeded any recorded for the

Fig. 60. Braeriach from Ben Macdhui, Cairngorms. The high granite boulder fields, montane heaths, corries and snowbeds, representing an outpost of Arctic habitats, the largest in Britain. Altitude 1190 m; date 15 June 1982. Haunt of Ptarmigan, Dotterel and Snow Bunting.

Arctic regions, and even in an average year Ptarmigan are at higher density in the Central Highlands than is usual in most parts of the World. Breeding density can vary quite widely within a small area, especially according to the extent of good dwarf shrub heath and abundance of rocks. There is, however, the same association as that found in Red Grouse, between lower average breeding density on the most infertile, acidic soils derived from 'poor' rocks such as granite, and higher density on the rather more fertile soils overlying 'better' rocks such as Dalradian and Moine schists. Good local densities on the Cairngorms granite are one pair to 3–4 ha, but on the Cairnwell–Glas Maol schists one pair to 1.2–2.0 ha have been consistently maintained, and here the population shows no tendency to cycle. Predominantly grassy ground, often associated with the 'best' rocks, such as limestone, is not good for Ptarmigan, and large areas of alpine grassland and moss heath with little shrub growth may be largely avoided.

The Ptarmigan is strongly territorial and the cock birds are at their noisiest in February and March, when they establish and defend their territories, croaking and challenging each other from prominent look-outs. A non-breeding surplus displaced by the territorial birds persists when the breeding population is settled to nesting. Eggs are not laid until at least May; the precise time varies according to the openness of the weather and state of vegetation growth. The well-camouflaged female sits tightly, and will often allow a close inspection without moving. There can be differences in clutch size between years, with a tendency to large clutches in years of population increase and initial peaks, and smaller sets during late peaks and subsequent declines. Breeding success varied from year to year, but also showed consistent differences between separate areas. On the granite Cairngorms, the numbers of young reared per adult is often 0.3 or less, but in the more fertile schist hills of the Glas Maol area it is seldom less than 0.7. Brood size showed the same tendency as clutch size in relation to point in the population cycle. Chick mortality showed surprisingly little relation to the normal vagaries of summer weather, though occasionally late snowfalls have a disastrous effect on breeding success. As in grouse, experimental work points to some intrinsic quality of the egg as the main factor affecting survival of the young.

The young of Ptarmigan develop the power of flight at a tender age, as early as 10 days old. At first, while the anxious female scuttles around in frenzied distraction display, the chicks usually crouch; but if one is discovered, and starts cheeping, it is a never-failing surprise to see the whole brood lift up and take wing to some safe distance. This seems to be a protective adaptation against ground predators which can easily run over the bare terrain. Predation appears to be the main cause of mortality, especially in adults, and Ptarmigan locally form quite an important food item for Golden Eagle, Peregrine and hill Fox. The chicks are also vulnerable to

Stoats and Weasels, which can range quite high. Watson (1965) nevertheless does not regard predation as an important factor in limiting population at any time of year.

The late summer maximum population usually shows little decrease before the following spring, for the bird is little shot nowadays. Ptarmigan flock as winter approaches, and when snow deeply clothes the high ground, these flocks often move downhill, sometimes well below the normal lower breeding limit. Significant interaction with Red Grouse is rare, for these stay in separate flocks and usually also move to still lower levels. The Ptarmigan, by this time in their winter white, often rest and roost in hollows in the snow, but seem not to tunnel into it as do their kind in the far north. As spring approaches, they move back and pair off to take up territories, becoming more aggressive to all others except their mates. This is when there appears to be a displacement of birds that are unable to compete and hold territory. The non-territorial birds can for a time replenish losses in the territory holding pairs, but they thin out and mostly disappear as the nesting season advances. Their fate is unknown, but they evidently move to lower ground or more marginal hills farther away, and there suffer increased mortality. This self-induced spring decrease may vary from 18 to 47% of winter numbers, suggesting that the breeding stock is limited by the birds' own behaviour, not by survival from the end of the previous nesting season.

While regarded by many writers as a relict species in Britain, the Ptarmigan is a pretty successful one, in that it occurs in most parts of the Highlands where there is suitable habitat. Given some local decline or retreat, and present absences from certain outlying hills, there is a large and extensive population which probably averages 10 000 pairs. Although there appear to be some similarities between its ecology and that of the Red Grouse, there is no evidence that Ptarmigan have shown any long-term decline in numbers within their main range during this century.

All the remaining high mountain birds are much rarer, apart from being mostly summer visitors. Even the next most abundant, the DOTTEREL, is a fringe species with us, for there is a large extent of suitable habitat where it does not occur. It nowadays nests sparingly in the Austrian Alps and Italian Appennines, and there were formerly sporadic nesting places in the Carpathians and Riesengebirge. It could thus be termed Arctic-Alpine, but is mainly an Arctic species. Dotterel nest widely and in some numbers throughout the mountain ranges of Scandinavia, and in the far north of Europe and Asia they are widespread on dry tundras down almost to sea level, though less continuously distributed than some northern waders. Although once recorded breeding in Alaska, it has not otherwise succeeded in colonising the New World.

The earlier history of the Dotterel as a British breeding bird rested mainly on observations in the Lake District and adjoining Pennines (Macpherson, 1892), but its main stronghold proved to be the central Grampians and Cairngorms (Harvie-Brown and Buckley, 1895). It is unclear how many Dotterel once bred in northern England, but up to around 1880 there could have been at least 50 pairs in good years. Declines were reported by 1900 but, until 1926, 5–10 pairs probably bred in most years. The end to regular nesting here seemed to be 1927, and from then until 1960, only about 8 nests were reported in total. Nesting has again been more regular during the past quarter-century, and spread over at least 10 different hills, but 4 pairs in 1979 is the highest known total, and in 1985 there appeared to be none. The statement by Blezard *et al.* (1943) is still a neat commentary on the status of the bird in the Lakeland faunal area.

'In part, the elusiveness of the Dotterel is accounted for by the fact that any given haunt is not necessarily resorted to annually, and may not be occupied for several seasons. It is sufficient to add that, in one haunt or another, the Dotterel continues as a nesting species, though a scarce one'.

The history of the Dotterel in the Southern Uplands is still less clear. Robert Service reported that it bred sparingly on the higher hills of Galloway and the Moffat–Tweedsmuir ranges. There was a dearth of records from 1900 to 1965, but during the past two decades up to 12 known nestings have occurred, on almost as many different hills. The first authenticated breeding of the Dotterel in Wales was not made until 1967, when R. Goodier discovered a nest on the Snowdonian mountains. Nesting appeared to continue in this locality for a few years more but has lapsed again. There is one recent nesting record in Ireland (1975).

Much of our knowledge of the Dotterel as a British breeding bird comes from many years of dedicated field work, often under the most arduous

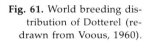

Fig. 61. World breeding distribution of Dotterel (redrawn from Voous, 1960).

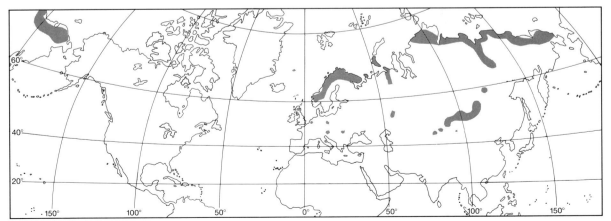

conditions, by Desmond Nethersole-Thompson. In his monograph (Nethersole-Thompson, 1973), he surmises that the recent increase of Dotterel in marginal districts, both to the south and to the north of its main breeding region in the central Highlands, probably reflects the post-1960 trend to appreciably colder spring weather. There has been a marked increase in the frequency and size of passage 'trips' of Dotterel on hills in northern England during the past 20 years or so. While this is matched by an increase in breeding within the region since 1960, most of these birds move on elsewhere, but their eventual destination is unknown. The Victorian hey-day of the Dotterel in northern England was associated not only with large hill trips, but also with regular parties in the adjoining lowlands. The breeding decline here was blamed upon the slaughter of both passage and nesting birds for their feathers, valued in the making of trout flies. This vicious practice had, however, ceased by about 1900, after which the bird continued to nest sparingly but regularly for another 25 years. Irregular nesting thereafter coincided with infrequency of spring passage records.

Nethersole-Thompson (1973) believed that the Scottish breeding population fluctuated markedly between 1930 and 1970, reaching at least 150 pairs in peak years. His figures show substantial year to year changes in numbers on his main study areas in the central Grampians and Cairngorms, and other observations suggested similar annual or periodic variations in other Highland breeding areas. The first attempt at a comprehensive survey of the British breeding population of Dotterel was made in 1987 by Des Thompson and Hector Galbraith. It produced the astonishing figure of over 500 pairs in the Highlands, but only 2 pairs each in northern England and southern Scotland. This has led to controversy, with Watson and Rae (1987) claiming that previous estimates of population have been far too low, and that annual numbers were probably always nearer this level in the Highlands.

The evidence suggests strongly to me that there were, indeed, many more Dotterel nesting in Scotland in 1987 than during the period 1930–70. Not only was there a virtual doubling of breeders, compared with previous peak numbers, on certain hills very carefully worked in the earlier period, but Dotterel bred on many hills with no previous nesting records. Some of these hills had been visited sufficiently often by bird watchers since 1955 to make it unlikely that Dotterel had simply been undetected there. Sample counts suggested that 1986 may have been as good as 1987, and numbers may have built up over some years. High numbers were maintained in 1988, and in 1989 a total of 867 pairs was estimated (D. Thompson). The increase in the central Highlands would also match the tendency to colder springs since 1960, though only in a general way over this period. Only careful monitoring into the future can chart the dynamics of our Dotterel population, and its possible connections with climate.

This would be a considerable undertaking, for the species is a most tantalising subject for study. The nesting habitats are both distinctive and circumscribed, but the combination of bad weather and uncertainty in the birds' behaviour can make accurate counting of nesting Dotterel a demanding enterprise. Perhaps the best time for assessing numbers is in the few days before the eggs are laid, for the pairs which are going to nest are then usually dispersed on their prospective breeding areas and most visible, with often song and display. Once the eggs are laid, the female seldom goes near the nest, and finding this depends largely on the behaviour of the incubating male. Most Dotterel nests that are found are revealed after the sitter has left the nest, been detected running around the vicinity, and then watched back to the eggs. These are the easy ones, but during any search, for every bird which behaves so obligingly, there will probably be at least another which sits until almost trodden on. Occasionally, one of these tight sitters is flushed by the good luck that the line of walk is sufficiently close, but most are missed. And some of the 'runners' or 'fliers' can be shy and reluctant to return to the nest.

Most Dotterel with small chicks show themselves, running around and often giving distraction display, but they do not all hatch at the same time, some may have failed, and the young may be fairly quickly moved to different parts of the hill. The birds may also be counted at the end of the

Fig. 62. Dotterel at nest in stony turf of high mountain plateau in northern England. Despite the large numbers now breeding in the highlands, few pairs stay to nest in hill regions farther south. Derek Ratcliffe.

nesting season, when both adults and flying young have begun to band together again in parties prior to departure. But it cannot then be assumed that all the adults have bred, nor even – unless they have been ringed – that they are all necessarily the birds which nested in the immediate area. The only sure method of counting breeding Dotterel is to follow a nesting area right through the breeding season. At least several visits between pair separation from parties and fledging of young are needed to give a dependable picture, but a ringing programme is necessary to give an accurate interpretation of relationships between birds in flocks at the beginning and end of the nesting season. Recent ringing is revealing fascinating insights about hitherto unsuspected movements of Dotterel between different hills, and even between Scotland and Norway (D. Thompson).

Dotterel nesting habitat includes some of the bleakest terrain in our country. The typical haunts are the windswept summit plateaux or broad ridges and upper spurs of the higher hills, where short montane grassland, woolly fringe moss carpets, lichen heath, or open stony fell-fields cover most of the ground. Periglacial features associated with frost–thaw movements of soil and stone are often well developed. Smaller numbers nest on the flattened mats of heather or mixed montane dwarf shrub heaths so much favoured by Ptarmigan, at a slightly lower level. On the high Cairngorms, the gravelly, boulder-strewn barrens with *Juncus trifidus* tussocks and patches of lichen and fringe moss are the usual habitat. The ground is either almost level or with a slope usually not exceeding 10°. Winter snow cover is usually thin and soon disappears, but on the higher Scottish mountains there may be numerous large snow-beds in sheltered hollows nearby, when the Dotterel begin to nest.

In the Grampians and Cairngorms, most Dotterel nest between 900 and 1200 m. Smaller numbers breed down to 800 m, while Adam Watson found a brood of small chicks almost on the summit of Ben Macdhui at 1300 m. Interestingly, the Dotterel shows the same tendency as the Ptarmigan to follow its habitat downwards in a north-westerly direction. Regular nesting haunts in Ross-shire are at 850–1000 m, but in 1987 a group of 7 pairs was at 700–735 m, and a 1971 nest was at only 460 m on a windswept lower spur of high hills. Recent nesting records for Sutherland were at 520–920 m, but one was at only 350 m, though on terrain of the right kind, with prostrate heather carpets and much bare stone or soil. A recent Argyll nest was at only 450 m.

Surprisingly, though, the Dotterel nested most consistently at lower elevations at its southern breeding limits, in northern England. While most nests were at 800–900 m, a much favoured hill in earlier years was at only 725 m, and several nests were seen at around 600 m. Recent nests were mostly at 750–900 m, but two were at 640 m and 550 m. In this region, *Racomitrium* heath was once extensive, and preferred habitat, but some nesting haunts had a short, stony turf with the right general appearance, rather than the real Highland Dotterel habitat.

It is disappointing, and puzzling, that the recent population expansion in the Highlands has not so far extended to northern England, Wales and the Southern Uplands. Passage numbers are quite high in northern England, but few birds are attracted to stay, and it seems that there has been some kind of environmental deterioration during recent decades. The enormous increase in hill walking, reflected in the beaten paths across the hills, is an obvious factor, but Watson and Rae (1987) state that the much increased recreational pressure has not yet had a detectable effect on Dotterel numbers in the Cairngorms. On one fairly popular mountain top, several nests have been found within 10 m of the main track, even though the birds have a large area to choose from. Another change, which I have witnessed during my own life-time, has been the steady loss of fringe moss cover on the high tops of these southern hills, evidently reflecting the cumulative trampling, manuring and grazing effects of increasing numbers of sheep. Swards of fine-leaved grasses – sheep's fescue, wavy hair and bent – have gradually replaced the moss. The problem may be more the risk of trampling on nests than the vegetation change itself. My NCC colleagues are studying the problem and suggest that acid rain could have contributed to these changes, which may involve food supply. Predation is another possible factor.

Fig. 63. Visitors to the high tops: Dotterel, Swifts and hikers.

Dotterel are seldom seen much below the level of their high nesting grounds, once breeding has begun. Largely invertebrate feeders, they forage on the dry habitats of their nesting area, but also resort to marshy ground and wet flushes in the immediate vicinity. Their reversed sex roles usually leave the male with the bulk of parental duties, to the point where the eggs are unattended while he goes off to feed. A bird disturbed from the nest usually takes the opportunity to pick repeatedly at ground insects and other small creatures as it runs around or works back to the eggs. It is a mystery how birds such as this re-find their nest: they may be up to several hundred metres from it, and they run all the way back, with their eyes no more than 15 cm above the ground, yet they know precisely where the eggs are, and return unerringly to them. This is homing ability with a very fine tuning. Some of these hill birds, such as Dotterel, also seem to be able to see through mist.

The Dotterel has long been known as a bird in which the normal roles of the sexes are largely reversed. The more brightly coloured female performs song displays and courts the male, and after she has laid, her mate takes over incubation and care of the young. To this is sometimes added the further departure of polyandry, in which the female takes a second mate and lays a second clutch for him to hatch (Nethersole-Thompson, 1973). Females seem usually to take little further part in family duties, though they appear again in the mixed flocks of old and young at the end of the season. Hatching success is good in many areas, but chick mortality seems often to be rather high, and Watson and Rae (1987) report from 0.2 to 1.2 (mean 0.5) young reared per adult in end of season flocks. Both breeding density and production of young were found to be consistently better on areas of schist with more fertile soils than on highly acidic granite.

Breeding density varies widely; some hills have always been especially favoured. Nethersole-Thompson (1973) reported a maximum density of 11 pairs in 165 ha (6.7 pairs/km^2) of suitable habitat on the Central Grampians in 1953, but this has been far exceeded in recent years (D. Thompson, personal communication). While nests are sometimes fairly regularly spaced, the Dotterel does not appear to be normally territorial, and nests can occur in quite dense clusters. This is an unorthodox bird which is still yielding up fresh secrets, and we must look forward to their publication. It is one of a group of plovers which have adapted to a range of mainly bare, sandy and stony habitats across the World, and has interesting relatives nesting in contrasting lowland habitats in Australia and New Zealand.

The SNOW BUNTING is the last of the trio of regular high mountain breeders in Britain, but by far the rarest and most elusive. This is a true Arctic species, with a virtually continuous circumpolar distribution. It breeds up to latitude 83° N and has been seen in the polar wilderness at

87° N, but migrates southwards at the onset of the Arctic winter to seek more congenial quarters in the temperate zone, and so becomes a familiar visitor to this country from October onwards. It has two main habitats then: sandy shores, mainly along the east coast of Scotland and England; and inland hill country. Our total wintering population, derived from northern Europe, Greenland and even North America, is then estimated at 10 000–15 000 birds (R. Lambert, in Lack, 1986). Towards their southern limits of breeding distribution in regions with milder climates, e.g. Iceland and southern Scandinavia, Snow Bunting populations appear to be more sedentary.

When this wintering population moves back north, during April and May, flocks drifting across the Highland mountain ranges are a frequent sight, and they travel through much apparently suitable nesting ground. Yet, by the end of May, they have nearly all gone, impelled by that urge which makes them home back to the vicinity of their birthplace beyond our islands. A mere handful of pairs stays behind to nest, and only in the Scottish Highlands. We are again indebted to Desmond Nethersole-Thompson for much of current knowledge of the species as a breeding bird in Britain.

Since records were kept, the high montane ground of the Cairngorms has always been the summer headquarters of the Snow Bunting here. Breeding has occurred on many other high mountains scattered across the Highlands, but apparently seldom if ever with any regularity in any one locality. In many years, only the Cairngorms are known to have held nesting Snow Buntings, and even this stronghold has occasionally shown a complete blank. Even allowing that only a small fraction of the total extent of suitable ground in the Highlands has ever been searched in one year,

Fig. 64. World breeding distribution of Snow Buntings (from Voous, 1960).

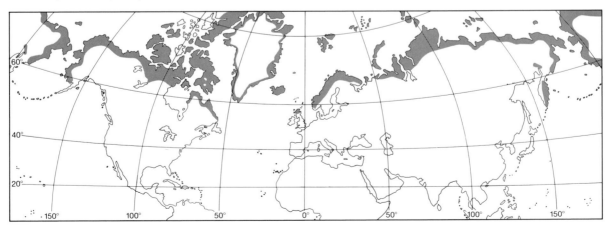

especially before 1900, this is truly a fringe species in Britain. During this century, 5 pairs has been regarded as a good annual total for Scotland, with perhaps 10 as a maximum. The past two years 1988 and 1989 have, however, produced the extraordinary totals of 60–80 and 50–70 breeding pairs (D. Thompson, unpublished) to set us wondering what the future holds.

The summer haunts of Snow Buntings in the Highlands are high corries and upper slopes with block screes, boulder fields and rock outcrops: places where snow patches linger in sheltered spots until late in summer,

Fig. 65. Breeding distribution of Snow Buntings in Britain and Ireland during 1968–72 (from Sharrock, 1976).

or even last the whole year through. Nethersole-Thompson (1966) has pointed out the preference of Snow Buntings for mountains with the longest-lasting snowbeds, and the frequent proximity of nests to these winter relics. The nest sites are in deep and protected crannies among the rocks. Almost identical habitats occur in breeding haunts on the Norwegian fjells, such as the Jotunheim and Hardangervidda, though on a vastly larger scale, merging into permanent snow- and ice-fields, and carrying extensive, large populations of Snow Buntings. In the Arctic this kind of terrain occurs down to sea level, and the bird with it, but occupying a much wider range of habitats. Haviland (1926) found the species nesting in swamps and low ground, especially near heaps of driftwood at the mouth of the Yenisei River in Siberia. In other parts of the Arctic nests are in a wide variety of situations, usually amongst loose rocks, but also in cliffs, banks and walls. Human settlements in the far north are much frequented, and nesting places among dwellings are commonly found.

The wider choice of habitat elsewhere gives some credibility to occasional past reports of nesting on seacliffs and low rocky hills close to the north-west Highland seaboard, or in the Islands, including Shetland and St Kilda. Yet it is mainly the classic high montane habitats which draw the Snow Buntings that remain to breed with us. This suggests that these birds come mainly from the population which is adapted to the most similar

Fig. 66. A fringe species from the Arctic: Snow Buntings in a high corrie.

kind of ground, and experiences the same day-length regime, i.e. that of south-west Norway. The next nearest large population is that of Iceland, which is regarded as a separate race *insulae* and distinguished by dark-rumped males, as distinct from white-rumped males in the typical race *nivalis* from Scandinavia and Greenland.

While it is possible that small groups of Scottish breeders could be self-sustaining for short periods, Nethersole-Thompson believes that the fringe populations of Highland Buntings depend on periodic immigrant renewal of the stock. Since dark- and white-rumped cocks occur in roughly equal numbers in our summering population, he suggests that this is derived from both Icelandic and southern Scandinavian populations. Although these are largely sedentary, small numbers from them undoubtedly reach Britain, and probably especially during severe winters in their home countries. Since migratory urge is weakest in these birds, the chances of some of them staying here the following spring may be fairly high. It is uncertain whether the small British breeding population is resident, partial migrant or migrant, for we do not know where these birds go in winter. Since Snow Buntings regularly come to skiing areas to eat discarded human food, Des Thompson suggests that the Cairngorm birds need winter no farther afield than the ski grounds in Coire Cas.

Snow Buntings have a mixed diet, of both invertebrates and vegetable matter. Insects and other small animals are probably the main food in spring and summer, and buds, fruits and seeds during autumn and winter. Those birds wintering amongst the hills feed mainly on the seeds of grasses and rushes. Nethersole-Thompson and others have described how in summer Snow Buntings often seek insect prey on or around the edges of late snow patches, and these habitats may facilitate feeding. Yet Milsom and Watson (1984) found no correlation in the Cairngorms between number of cocks on a territory and extent of snow cover in late spring. It seems most unlikely that scarcity of late snow cover or availability of food more generally limits the numbers of Snow Buntings summering in the Highlands. This is, indeed, a good example of a species occupying only a tiny part of the available habitat, and the reasons for this limitation are to be sought elsewhere.

Although Scottish breeding Snow Buntings go through the motions of territorial behaviour, this can hardly have the same significance as in large and dense breeding populations. With neighbouring nests averaging 1500 m apart, and some nests completely isolated, Nethersole-Thompson found that such behaviour by the males was related more to seeking and holding of a mate than to defence of an area. In the Cairngorms, it was usual for there to be surplus, unpaired cock Buntings, which continued to sing and display, often ranging over large areas, while paired birds had eggs or young. These spare cocks would quickly pay court to paired hens if opportunity allowed, and could be most persistent suitors even when repeatedly driven away by the established males. In 1972, 5 nesting pairs and 7–8 unattached males were known in the Cairngorms. Males are thus evidently more readily motivated than females to remain to breed in the Highlands, but perhaps this merely reflects the males' stronger territorial urge. Polygamy is, however, occasional in Cairngorm Snow Buntings, and females have mated with previously unattached males to produce a second brood while their original mates were tending the first. Orthodox second broods are also quite frequent in Scottish Buntings. Overall breeding performance of this fringe population has been quite good, but there are no data on subsequent mortality of either young or adults.

Nethersole-Thompson (1976) believes that the improved summer status of the Snow Bunting in Scotland, when the period 1920–60 is compared with that since 1960, reflects the cooling of the climate, at least during spring, over this time. Exceptionally severe individual winters, such as 1947, have produced good bunting years, but the best numbers and regular nesting have been since 1960. This increase may owe as much to more inviting springtime conditions on the Highland mountains, as to an extra winter displacement of birds from Iceland and Scandinavia. The exceptionally heavy winter snowfalls and prolonged spring snow cover of the years 1989–90 in Scandinavia – a perverse effect of mild winters with increased precipitation – could also have displaced more of the high level birds to the Highlands.

During the past 15 years, new northern species have been added to our list of high montane nesting birds, mainly in the central Highlands. Shore Larks were first, in 1972–73 (Campbell *et al.*, 1974), followed by Lapland Buntings in 1977 (Cumming, 1979); both are species of high-level grasslands and heaths. A long-awaited 'first' was the discovery of breeding Purple Sandpipers in 1978, on bleak and stony fell-field terrain (Dennis, 1983). Single birds of this species, seen in May of 1971 and 1975 on the same hill in northern England, could have had sitting mates but no evidence of nesting was found. Shore Larks and Lapland Buntings have bred subsequently though erratically, but Purple Sandpipers seem well established in at least one locality, and have probably nested in at least two others far distant. Only a few pairs in any one year have been found, but there is a vast extent of suitable ground through the Highlands, much of it searched only sporadically.

This last reflection does, indeed, raise the tempting question of whether careful and systematic survey of the high tops would bring to light not only additional nesting pairs of these species but also further new discoveries. The possibilities are always there, or so one likes to think. Pairs of Sanderlings have twice been seen on high fell-fields in nesting time, though this is a high Arctic nesting species, breeding no closer to us than east Greenland.

Single Snowy Owls have summered on the high Cairngorms, and probably only a scarcity of food inhibits nesting attempts on the montane plateau, which more resembles the typical Scandinavian breeding grounds than the actual nesting place of Fetlar (Fig. 3). One in 1987 developed an unfortunate preference for young Dotterel and severely reduced breeding output of this bird over a fairly wide area (D. Thompson, personal communication).

Our descriptive scan of the mountain and upland birds appropriately ends here, on the high tops. It remains to consider further some of the influences which rule their lives, and to think of the future trend of events, including both those within and those beyond our control.

8

Geographical aspects of the upland bird fauna

This chapter considers some further questions about the composition, distribution and status of our upland bird fauna, especially in relation to the geographical limits of species' ranges.

Present composition

The history of post-glacial change, involving geographical shifts of main habitats and their associated bird assemblages, helps to explain the particular assortment of species occurring on our uplands today. The retreat of one bird group and its replacement by another has matched the migrations of vegetation zones across these islands and up their mountains, in response to climatic change (see Introduction). In Table 1, I have assigned each upland species to a broad geographical category, based on World breeding distribution. As one might expect, geographical and habitat distribution of upland birds in Britain and Ireland tend to be a finer scale reflection of the global patterns.

The Arctic–Boreal bird fauna has become depleted in Britain, and even more in Ireland, with some remaining species acquiring a relict or fringe status. The group especially adapted to the vast Arctic tundra regions and the northern mountains is only poorly represented in these islands, where climate is now so marginal for it, and equivalent habitat so restricted. The Scandinavian montane and Arctic dry tundra species which breed in Britain are virtually all montane here; of these only the Ptarmigan occupies most of the suitable habitat available. Despite the recent increase of Dotterel (the only montane species known to breed in Ireland), there is much apparently suitable ground where it is absent or only sparse. For the other montane species, conditions on our mountains are evidently only marginal at present: the Snow Bunting, Lapland Bunting, Shore Lark, Purple Sandpiper and Snowy Owl either have tiny populations or only nest erratically.

The more widespread Arctic or Subarctic to Boreal group which occurs on both treeless tundras and in open forest habitats is better represented here, and also tends to have a mainly sub-montane distribution. The Golden Plover, Dunlin, Merlin and Ring Ouzel probably occupy most of the suitable habitat in Britain, or did so until recently. This is less true in Ireland, though all four are fairly widespread there. The Greenshank, Black-throated Diver and Red-throated Diver are widely distributed over suitable terrain within the Highlands and Islands, but are most numerous in the north and west and have a fairly rigid southern limit. Yet several species of the taiga marshes and lakes, and the scrub tundra, the Whimbrel, Red-necked Phalarope, Wood Sandpiper, Temminck's Stint, Ruff and Common Scoter, are fringe species here. Except perhaps for the Phalarope, shortage of suitable habitat is unlikely to be the limiting factor. The species most widespread in our uplands, especially within the submontane zone, tend to be those whose total distribution extends from the Boreal to the Temperate regions or even farther south. And several of our notable opportunists, such as Peregrine, Kestrel, Buzzard, Golden Eagle, Raven and Cuckoo, have an extremely wide global distribution.

Interestingly, our upland bird fauna has little affinity with that of the Alpine regions. Only 3 species are Arctic–Alpine (Ptarmigan, Dotterel, Ring Ouzel) and we have none of the strictly Alpine species: Alpine Accentor, Water Pipit, Rock Thrush, Blue Rock Thrush, Wall Creeper, Snow Finch and Alpine Chough. These birds mostly live in dry, steep and rocky terrain, so widespread in the sharply contoured Alps, and their adaptation contrasts with the waders, so strongly represented on the extensive wet and dry tundras and fell-fields of the Arctic regions. We have an abundance of both types of mountain habitat, and the strong emphasis of northern, tundra-breeding species in Britain and Ireland suggests that our upland avifauna owes its origin mainly to southwards displacement of bird assemblages from the Arctic and Boreal regions during the last glaciation. Britain and Ireland have acquired a few strictly Alpine plants (e.g. Spring Gentian) at some point during Quaternary history, but if any of the birds peculiar to southern European mountain regions reached us, they died out here long ago.

A few of our upland species have a peculiar, discontinuous (disjunct) World distribution, e.g. Shore Lark, Twite, Dipper, Greylag Goose and, above all, Great Skua. The historical and evolutionary significance of these patterns is discussed by Voous (1960).

Island effects

Island bird faunas tend to show decrease in number of species with reduction in size of area and distance from the nearest large land mass: the area and isolation effect (MacArthur and Wilson, 1967). The upland bird faunas of the islands of Britain show something of this tendency, though there is

Table 4. *Island distribution of upland birds*

Island (Area in km²)	Ireland (84 430)	Lewis & Harris (1994)	Skye (1667)	Shetland (1426)	Islay (1034)	Orkney (975)	Mull (914)	Uists & Benbecula (756)	Isle of Man (588)	Arran (430)	Jura (365)	Rhum (107)	Tiree (75)	Coll (74)	Colonsay (41)	St Kilda (8.5)	Scottish Highlands mainland (c. 45 000)
Number of species	48	44	47	38	45	42	41	43	27	43	44	35	25	25	30	14	65

Species mainly in the Scottish islands, though only on a few: Whimbrel, Red-necked Phalarope, Great Skua, Arctic Skua, Snowy Owl.
Species with a patchy distribution in the Scottish islands: Black-throated Diver, Wigeon, Common Scoter, Greylag Goose, Hen Harrier, Stock Dove, Short-eared Owl, Nightjar, Dipper, Ring Ouzel, Grey Wagtail.
Species on very few of the Scottish islands: Goosander, Ptarmigan, Black Grouse, Dotterel, Greenshank.
Records mainly from Sharrock (1976) and Thom (1986).

not an appreciable fall-off in species variety until islands are less than *c.* 300 km² (Table 4). The central proposition to explain the effect is that the balance between the rates of extinction and of immigration changes adversely as area decreases and isolation increases. Reduction in habitat diversity and extent with decreasing size of area is, however, an important factor (Reed, 1981), though lack of suitable habitat accounts for surprisingly few absences of species from the islands listed in Table 4. Only when the island is very small, notably in the case of St Kilda, does this have much obvious significance. Other small islands not listed, e.g. Gigha, Eigg and Canna, have few upland birds because they have little high ground. The poor showing of the Isle of Man may reflect its southern position and lack of hill wetland habitats.

Yet the absence of some species from certain islands is puzzling, and hardly explicable by supposing that the birds concerned do not reach these places, for they are mostly mobile species if not actual migrants. The Isle of Skye has almost as many species as the whole of Ireland. It is especially surprising that Shetland lacks Hen Harrier, Short-eared owl, Golden Eagle and Black-throated Diver, while the Red-necked Phalarope is a doubtful nester on Orkney, and the Greenshank only a sporadic breeder in both these island groups. The Dotterel is not known to nest anywhere in the Scottish islands, there are only old records of Snow Bunting from Shetland and St Kilda, and the Ptarmigan is only on Skye, Mull and Arran.

It may be that the superficial appearance of suitable habitat is deceptive, and that for some absent species there are more subtle conditions which are not satisfied. The resistances to the spread of many birds are far from understood, and we shall look briefly at how some species have enlarged their range.

Relationships to climate

The uplands are only a summer breeding habitat for many birds: at least 47 out of 66 species move elsewhere after the nesting season, varying from local shift to adjoining lowlands or coasts, to full-scale migration to distant lands. This repeats the larger pattern of migration to more congenial wintering haunts farther south from Arctic breeding grounds, where only a few hardy species pass the whole year round. Our upland birds can afford to take their time over nesting, and have adjusted to the cool temperate climate. Whereas the rigours of the high Arctic climate compress the breeding season into some 10 weeks between appreciable spring snow melt and first autumnal snowfall, our upland birds usually have from 3 to 10 weeks longer. In the Arctic there is little chance of repeat nestings if the first attempts fail, whereas here, birds such as the Golden Plover will repeat at least twice. The continuous daylight of the Arctic summer allows concentrated bird activity, though in northern Scotland, midsummer night is only a few hours of twilight darkness. Most of our upland species are single-brooded, but the Stock Dove, Dipper, Ring Ouzel, Stonechat, Meadow Pipit and Pied Wagtail are commonly double-brooded, while Snow Bunting, Twite, Whinchat, Grey Wagtail, Skylark and Wren sometimes are.

The daily and seasonal variations in meterological conditions which we call weather have direct and immediate effects on birds. Although, as warm-blooded, feathered creatures, birds are somewhat buffered against cold, their mobility allows them to respond quickly to unfavourable temperatures. Severe conditions, with heavy snowfall and/or prolonged frost, often delay the return of many species to the hills, and the onset of nesting. Snow can fall on the high tops up to late June, and sometimes heavy falls in late April or early May wipe out most nests of earlier breeders such as Red Grouse and Golden Plover, causing them to repeat and have a late fledging of their young. Sometimes, in years of bad weather, various species have a reduced final output of young, compared with more normal years. Even the Peregrine has suffered appreciably higher than average loss of eggs and mortality of nestlings during several unusually wet and cold springs during the past 20 years (especially in 1981). After the Arctic winter of 1962–63, upland breeding populations of Lapwing, Redshank and

Stonechat appeared much reduced in many areas, though Golden Plover were less noticeably affected.

Usually, these adverse effects of weather are only temporary set-backs, and milder years allow numbers to build up again. Some species, especially the larger birds of prey, show a remarkable constancy in breeding population in some areas. A few, notably the three grouse species, have large and sometimes cyclical population swings. In between these extremes, most upland birds show some degree of fluctuation over a period of years.

I am more concerned here with *long-term* population trends in relation to shifts in climate. It has been noted that the tendency to cold, backward conditions, especially in March–April, since around 1960, has been matched by an increase in the number of Arctic–Boreal species nesting in Britain (Sharrock, 1976; Williamson, 1975). The appearance of Wood Sandpiper, Purple Sandpiper, Shore Lark, Lapland Bunting, Snowy Owl, Great Northern Diver, Redwing, Fieldfare and Bluethroat as nesting species in Britain, and the return to regular breeding of Snow Bunting in the Highlands and Dotterel in northern England, all fit quite neatly with the evidence for a slight shift to a more arctic climate. On the other hand, the Whimbrel, Red-necked Phalarope, Greenshank, Red-throated Diver and Black-throated Diver have not responded by a southwards shift in breeding limits.

Moreover, while one would expect a corresponding retreat of birds adapted to warm climates, there is rather little evidence of this. The disappearance of the Woodlark from the lower uplands and marginal lands of Wales seems to be the clearest example of retreat under deteriorating climate, with the 1962–63 winter as the extreme which applied the *coup de grâce* (Condry, 1981). The decline of the Nightjar and Red-backed Shrike are far less convincingly the result of climatic shift, and frost-sensitive species such as Stonechat and Dartford Warbler continue to show decline in response to cold winters, but then recover again over a run of better years.

How would climate control of upland bird distribution limits and numbers actually operate? Presumably, for species already breeding in Britain, one way would be by changing the balance between recruitment and mortality, if only at the edge of the range. Increase in population will tend to be followed by spread, and vice versa. For species to colonise or recolonise Britain or Ireland involves a further step, either in the overspill of a more northerly population surplus, or a tendency for northern migrants to stay farther south than usual in order to breed. Harvie-Brown (1906) believed that an increase in nesting Dotterel in the southern Highlands around 1903–05 was associated with a southwards shift in migration pattern occasioned by a run of cold, backward springs. The recent increase in our Dotterel breeding population (pp. 188–189) similarly matches the considerable increase in spring passage 'trips' of Dotterel reported from both upland and lowland England during the past 10 years or so.

If severe weather delays migration or deflects the migration stream southwards, it may increase the chances that a few individuals will stay to nest, amongst the northern European migrant species which either winter here or pass through in spring. Perhaps there are effects of visual stimuli, such as the amount of snow-cover on our mountains when these migrants reach a particular point in the spring renewal of their breeding rhythm. Or perhaps the urge to nest overtakes the urge to migrate. (See also p. 198).

An unanswered question is how far our fringe upland species depend for survival on the productivity of their own small populations, or on replenishment from migrant stock. Most of our upland waders also have large migrant populations from northern Europe either passing through or wintering in Britain and Ireland. Yet it cannot be assumed, without good evidence, that these migrants help to 'top up' our native breeding populations. They may do so, if only in certain years, but the instinct to return to birthplace to breed is so strong that breeding populations of many species are likely to be effectively isolated, e.g. the southern and northern populations of Golden Plover. Snow Buntings reach Britain in large numbers every year, but Nethersole-Thompson (1966) believed that irregular breeding during 1900–65 was because small groups of pairs only became established occasionally and then tended to die out over a few years under the marginal conditions. Our Red-necked Phalarope populations are evidently isolated and doubtfully self-maintaining (p. 175).

Limiting conditions

One route to understanding plant and animal distribution is by identifying the limiting conditions for successful existence. On the broad geographical scale this is often done by measuring the conditions, especially of climate, which most precisely fit the map limits of distribution. Voous (1960) gave numerous correlations between limits of breeding range for European birds and specific isotherms of mean temperature for both the coldest and warmest months (January and July). Many others have made such correlations for particular species, and the *Atlas of Breeding Birds in Britain and Ireland* (Sharrock, 1976) provided a series of transparent overlay maps of various environmental factors, for readers to do so. Where a close 'fit' exists, limits of bird breeding distribution are presumed to be determined in some way by the factor concerned.

Sometimes, direct observation of a bird's biology can be readily connected with these broad correlations. Sharrock (1976) notes that the association between breeding distribution of the resident Stonechat and the milder winter climate of western and south-western areas accords with the fact that this species suffers heavy mortality in severe winters. As a small, insectivorous bird, it must suffer when frost and snow seal the land for long periods. Sometimes these map correlations are useful in suggesting lines of worthwhile study of ecological relationships. The drawing of ecological inferences from map correlations about factors which limit species' ranges is nevertheless fraught with problems, especially for animals. For those plants which appear to show a fairly direct physiological relationship with limiting climatic conditions, the hypothesis may be testable in controlled environment experiments. With birds this is very difficult, and the causal link in the correlation may be through the vegetation and food supply of its nesting and feeding habitat.

Where a bird species is adapted to a particular habitat, its distribution limits will tend to be circumscribed by whatever conditions restrict the occurrence of that habitat, but these conditions may vary over the total range. The northernmost or uppermost limits of tree growth (the tree-line) are closely correlated causally with a mean summer temperature of 10 °C during the two warmest months (Manley, 1952). Yet, under the extreme oceanic climate of northern Scotland and western Ireland, the tree-line is depressed markedly by excessive windiness, and disappears altogether over the large areas of wet blanket bog which occur down almost to sea level. This oceanic climate thus gives a much larger extent of suitable habitat for upland and northern birds than would seem likely from isotherm maps. The low-lying Flow Country is remarkable in having a breeding bird assemblage closely equivalent to that of Low Arctic wet tundras, though some species are represented by equivalent life-forms, e.g. Whimbrel by Curlew, northern geese by Greylag, and Willow Grouse by Red Grouse.

Another complication about inferences from environmental correlations with bird distribution is that the same ecological relationships do not always apply throughout the range. In species with a very wide World distribution, regional adaptations can develop, e.g. in the Shore Lark (Voous, 1960). These ecological divergences are likely to become genetically fixed. Some species may be undergoing a change of range: either a retreat from some adverse influence, such as a new competitor or loss of habitat or food source; or an expansion which is merely the crossing of a previous barrier or continuation of well-established advance, but may sometimes result from exploitation of a new ecological niche. Occasionally there appear to be new species in the making through isolation and divergence from the original stock.

Ecological divergence within species

The Red Grouse has diverged from the northern Willow Grouse into an insular southern form in Britain and Ireland. The Willow Grouse ranges from open forest to more montane habitats above the tree-line: the medium shrub and dwarf shrub zones, mixed with more open tundra. The medium shrub zone in Britain has been largely lost, but its derivatives represent the bottom of the Ptarmigan zone here. The Red Grouse is thus a more submontane form of the Willow Grouse which has adapted to our oceanic upland climate, with its dwarf shrub heaths, within and above the original forest zone. Under the warmer conditions, the Red Grouse has lost the seasonal alternation in plumage phases of its relative. The white plumage of Willow Grouse and Ptarmigan is an adjustment to prolonged winter snow cover in the haunts of these sedentary species. The Ptarmigan in Britain lives within the montane zone and retains its seasonal dimorphism, but the equally sedentary Red Grouse has moved farther downhill here, and snow cover is not long enough under the milder climate for white plumage to be advantageous in winter.

In both plumage and physiology, the Golden Plover at its southern limits on the British and Irish moorlands is a rather different creature from the Arctic–Subarctic bird. The two become indistinguishable during winter, so that their degree of intermingling or separation within these islands is then unknown. Yet they must have quite different day-length adaptations in spring, for many of the northern birds are still in these islands when most of our residents are well settled to nesting. The adaptive significance of the plumage differences between north and south in Golden Plover is unclear.

Most of our upland species show little or no difference in plumage compared with northern European populations. A great number have diverged ecologically, through relatively recent adaptation to the extensive dwarf shrub heaths and grasslands which replaced the original forest cover. While the Red Grouse and Golden Plover have taken particular advantage of these expanded submontane habitats, their plumage divergence from northern stock appears to pre-date the last glaciation.

Several Arctic/Subarctic–Boreal waders belong essentially to the broad northern taiga zone of conifer and birch forests. They need moist ground or open water for feeding, and are especially associated with the bogs, marshes and open heaths of these northern forests, or the wet low shrub tundras just beyond them: habitats either rare or absent in Britain and Ireland. The Greenshank has made a widespread and successful adaptation to treeless moorlands in the Highlands, and the Wood Sandpiper, Temminck's Stint and Ruff have also used treeless habitats in their more scattered or sporadic Scottish nestings. The Merlin in the Boreal and Subarctic regions also lives in open forest, forest edge and low shrub tundra habitats, and has adapted especially to the heather moors derived from forest. Other species to expand their distribution and numbers greatly through deforestation are the Curlew, Lapwing, Snipe, Wheatear, Whinchat, Stonechat, Meadow Pipit, Skylark, Twite, Ring Ouzel, Cuckoo, Raven, Buzzard,

Golden Eagle, Kestrel, Hen Harrier, Peregrine and Short-eared Owl.

The Curlew in Britain and Ireland has further expanded its breeding range off the hills and onto the agricultural lowlands, while the Redshank has moved upwards to become a locally common hill bird during this century. The Oystercatcher and Ringed Plover have gradually colonised river shingle and stony lake shores in the hill country, mainly in Scotland, and the former has taken also to nesting on both enclosed grassland and open moorland in places. Our Ringed Plovers have not yet adapted to high montane fell-fields which represent the equivalent nesting habitats often used by this bird in the Arctic. The Chough shows a westward extension of its European range mainly into coastal habitats, but in the west of Britain and Ireland, it has adapted again to upland terrain.

Probably most of our upland species show a similar breeding biology in Britain to that elsewhere, and their habitats here thus tend to be scaled-down equivalents of those over their much larger European, Eurasian or global ranges. Ptarmigan, Dotterel and Snow Bunting in Britain live in montane habitats which are the counterparts of the dwarf shrub, moss and lichen tundras and fell-fields occupied throughout their main ranges. The birds associated with lakes, rivers and wet tundras mostly seek the nearest equivalents here to the habitats they use elsewhere in the northern hemisphere. The Golden Eagle, Peregrine, Buzzard and Raven breed in a similar range of places to those held elsewhere.

Resistances to spread

Some birds are prevented from realising their capacity for spread by quite different factors, notably persecution in the case of birds of prey. We have seen how the Golden Eagle, Buzzard, Hen Harrier, Red Kite, Peregrine and Great Skua have increased and spread with the relaxation of persecution. The Peregrine's recent adaptation to ground and tree nesting sites is a good example of how scope for further expansion can grow under reduced persecution and increasing population pressure.

For some fringe species, climatic conditions are evidently too marginal in Britain or Ireland for them to increase and occupy more of the available habitat, though a few are slowly improving their position (e.g. Dotterel and Snow Bunting). Sometimes, suitable nesting habitat may abound, but appropriate food sources are lacking, as for the Snowy Owl on the Cairngorms. It would probably help this bird and perhaps others if we had more variety of small ground mammals, including lemmings. Some species may run up against heavy predation pressure, which inhibits any attempts to colonise new ground. Sometimes there may be competition from established occupants to resist expansion of other species into new areas or niches.

The lack of forest marsh and low shrub tundra here may discourage some Arctic–Boreal waders from settling. Northern Scandinavian birds that appear here in autumn or winter but do not stay to breed include Bar-tailed Godwit, Spotted Redshank, Jack Snipe and Little Stint. These species visit our shores in some numbers, yet ignore even the large areas of suitable habitat in southern Scandinavia. Others which breed not far away in Scandinavia appear here so rarely that it is hardly surprising they do not nest in Britain or Ireland, e.g. Gyr Falcon, Great Snipe and Broad-billed Sandpiper. And those high Arctic breeders which ignore even the tundras of Lapland and Iceland, such as the Grey Plover, Knot and Sanderling, are even less likely ever to stay to nest near their wintering haunts in our islands. The reasons for the localisation of many species within the Arctic regions are simply not understood.

The triers: occasional birds of the mountains

Here and there among our uplands are birds which normally belong to quite different habitats, mostly in the lowlands. These occurrences may be just a passing aberration that lasts no longer than the individuals concerned. Or they may involve enough pairs, over a large enough area and for long enough, to be considered a local habit. If the habit spreads to become an established adaptation to new ground for a large segment of the population, it may have considerable significance. Permanent adjustments to new habitat are one process whereby the gradual evolution of new species takes place.

The sea-going Manx Shearwater resorts to the land only to breed, and mostly to maritime islands. On Rhum there is a huge colony of over 100 000 pairs breeding up to 800 m on the highest mountains, and up to 3 km from the sea. Certain hills are composed of peridotite, which weathers readily to deep loamy soils in which the Shearwaters can dig their nest burrows. The harder granophyre and Torridonian sandstone elsewhere on Rhum give only thin or peaty soils and have no Shearwaters. This is an opportunistic adaptation to unusual geology, and the Manx Shearwater utilises whichever coastal breeding places more readily provide nesting sites.

Montagu's Harrier had eighteenth century upland nesting stations on the North York Moors and Yorkshire Pennines (Nelson, 1907) and on the moorlands of Northumberland and Durham (Evans, 1911). Grouse preserving ended these occurrences well before 1900. During 1940–55, Montagu's Harriers were again on the increase in Britain and extending into moorland habitats. My friend Bill Robinson found them nesting in young Sitka Spruce plantations on Dartmoor in 1945, in similar situations to those utilised by Hen Harriers in Scotland. They nested for several years on the grouse moors of Co Durham, but faded out some time after 1955. In Scotland, Eddie Blake found a pair of Montagu's Harriers nesting close to Hen Harriers on the low-lying grouse moors of Perthshire in 1952.

The Barn Owl takes to the hills in some areas, such as Galloway, where nest sites in crannies of the lower crags at 300 m or so have been occupied in several places. At least three were close to Peregrine eyries, and in one case fatally so, in the attempt to widen habitat. In the Border moorlands, a nest site in a rocky wooded glen was a safer choice in the move uphill. Magpies usually keep to the valley bottoms and woods, though a few straggle up the slopes, nesting in hawthorns and other trees. In the Sheffield area, with a particularly strong Magpie population, one moorland pair had gone farther by nesting in the sheer gritstone face of Burbage Rocks, probably as an overspill from the saturation of normal nesting sites in the immediate area. Rather few House Martins nest in natural cliffs in this country, and of these, most prefer coastal rocks. There are inland cliff sites in various places, nevertheless, and I noted a few pairs nesting at 600 m on Craig Rennet in the Angus hills. Sand Martins sometimes nest far up hill streams, in steep-cut banks of alluvial sand or morainic till in which they can burrow. Colonies are usually of only a few pairs each, but a big moraine bank in Glen Artney, Perthshire, had over a hundred nest holes.

Occasional hill-going pairs of Blackbirds breed in precisely the same kind of habitats as Ring Ouzels: drystone walls, rocky moorland gills and steep crags, up to nearly 600 m. The same haunts in other years may have the real Mountain Blackbirds, but relationships between the two have not been observed. In the Outer Hebrides the Ring Ouzel is almost absent, and its place is taken by the Song Thrush, which nests high up in the barren, rocky hills. I encountered one singing on the block screes of Clisham, the highest hill, at around 700 m. Which species came first, or how substitution took place, is unknown, but it is an intriguing question. The Mistle Thrush almost qualifies as an upland bird, for it is quite characteristic of the lower hill grassland and the moorland edge. It frequently nests in the storm-tossed trees planted for shelter around shepherds' dwellings on otherwise treeless sheep-walks. Some pairs take to the open hill, nesting on bare crags or in old quarries, sometimes at modest elevations and in bleak settings.

The Robin is occasionally found in bare, rocky places in the hills, apparently nesting. Grasshopper Warblers nest on lower hill ground in some areas, mainly in long heather or dense, tussocky grassland; and they appear readily in the rank vegetation which develops on recently afforested lower ground. The Reed Bunting has spread into some of these habitats in certain lower areas, and into most of the northern and western Isles of Scotland. Some other small passerines occur where hill woodland thins out into grassland or heath with scattered trees: Willow Warbler, Tree Pipit and Redstart. The last is one of a number of species (p. 87) which in places makes use of buildings on open hill ground for nesting sites. Our only woodland wader, the Woodcock, now and then leaves the cover of upland woods to nest on open hillsides among bracken, heather or bilberry. Shepherds and keepers occasionally find nests by chance, but there is no means of knowing their real frequency. During their return migration in March, Woodcock pass through the lower hills and moorlands in some numbers, and are often flushed from their daytime resting places. This is probably when they most often fall victim to hill Peregrines.

Two species which during the nineteenth century were partly upland birds have retreated from this habitat. The Osprey disappeared from Britain between 1917 and 1954 and since its return appears to have been exclusively a tree nester, resorting mainly to rather open woodland of Scots pine, and feeding on adjoining lakes and rivers. Yet the Ospreys of the north-west Highlands, which drew the rapacious attentions of the Victorian collectors, nested typically on 'sugar-loaf' rocks in lochs in bleak and almost treeless mountain and moorland settings (Harvie-Brown and Macpherson, 1904). Similar nesting places were known in Galloway. It is strange that the 'Fishing Eagle' has not lately returned to these hill lochs. Could this be not so much a failure to adapt from tree to rock nesting as an effect of poorer food supply at the present day? Acid rain has reduced or eliminated fish from some Galloway hill lochs, but the north-west Highlands are thought to be much less affected. In the Lake District, there was reputedly an eighteenth-century Osprey nesting place on the crags above Ullswater.

The other bird which has lost its moorland haunts is the Marsh Harrier. Nowadays, this raptor is regarded in Britain mainly as a bird of the lowland fens and marshes with extensive reed-beds. Its classic haunts are the Norfolk Broads and the great reed fens at Minsmere and Walberswick. Yet Marsh Harriers certainly nested on the moorlands of north and east Yorkshire up to the mid-nineteenth century, and also on lowland raised bogs at Thorne and Hatfield Moors, south of the Humber (Nelson, 1907). There are rather less sure records of breeding on the Border moorlands of north Cumberland, but Canon Tristam reported Marsh Harriers nesting in the moorland parish of Eglingham in Northumberland up to 1830. The disappearance of these moorland Marsh Harriers is attributable to the relentless onslaught of the grouse keepers, which also eliminated the more widespread Hen Harrier for many decades thereafter.

Occasionally, there are reversed departures from the norm in which mountain birds resort to lowland habitats quite out of character with their usual haunts. As isolated instances, single pairs of Golden Plover were found nesting on wet meadowland near Carlisle, on stony ground by a disused airfield in Lincolnshire, and on the grassland of Oswestry racecourse.

By far the most startling innovation of this kind was the establishment of a population of Dotterel in the Dutch polders of east Flevoland from 1961 onwards. Several nests were found in some years, all on cultivated fields with young crops of sugar beet, potatoes, flax, wheat and peas, at elevations down to 4 m below sea level. It is not known how numerous the bird became in these Netherland habitats, and it seems that the polder Dotterel may have disappeared. Between 1961 and 1968 only 16 young hatched out of 29 eggs in 10 nests: a hatching success of 55%, compared with 92% for a sample of Scottish clutches (Nethersole-Thompson, 1973). The scales may have been weighted against this bold attempt to occupy new habitat. Yet this kind of event, amounting almost to an ecological mutation, may occasionally succeed and set in train the divergence of a species into a new evolutionary line. Occasional nesting is also reported from other lowland migration staging posts for Dotterel in Europe (Glutz, Bauer and Bezzel, 1975). Surprisingly, there are so far no records of breeding in the East Anglian Breckland, where on migration the species encounters flint-strewn heaths of short grass and lichen superficially resembling its high mountain nesting grounds.

Conclusions

Evolution is so slow a process as to be seldom directly observable. Even the modest steps towards new subspecies taken by our Red Grouse and, less distinctly, Golden Plover, are on a time-scale of tens of thousands of years. We can witness only the first steps, of change in distribution and/or habitat preferences, which may set a species on the road to divergence. Major geological changes are mostly too slow to produce noticeable effects on bird distribution, but response to climatic fluctuation is more visible within our life-span. Probably most of the observable changes in bird abundance and distribution within the past few centuries have been the result of human intervention. A few species at any one time also seem to undergo spontaneous and unpredictable changes in habits and distribution without the obvious stimulus of environmental change.

What does all this hold for the possibility of further exciting additions to our upland breeding bird fauna? Much will depend on the behaviour of climate, and particularly a holding to the recent tendency towards colder springs. Turnstones are strongly rumoured to have bred already, and Rough-legged Buzzard, Long-tailed Duck and Long-tailed Skua all seem possibles. Yet the resistances to the spread of species, or the return of those long-lost, are such that we may dream largely in vain. We may be able to help some to reappear, or even attract others afresh, but our concern will have to be mainly with those already here. Their needs, and our intentions about these, are the subject of the last chapter.

9

Conservation of upland birds

The problems

The conservation of Nature is concerned with the effects of human activities, which have become increasingly dominant during the past 3000 years. Man's destruction of the upland forests, and their replacement by grazing range, must have led to a great increase in open-ground birds, and perhaps to the development of new bird communities. Subsequent run-down in carrying capacity may have caused local decline in certain species, but some birds have benefited from human occupation of the uplands. Divergence into management for the three key animals, sheep, Red Deer and Red Grouse, has had variable effects. Sheep-walks have, on the whole, lost much of their dwarf shrub heath and become grassland- and bracken-dominated, but have provided a good food source for carrion-feeding scavengers. Grouse moors have remained heathery, and valuable habitat for many species, but the predators have generally been persecuted. Deer forests include some of the least modified of all our mountain country, but some have come to resemble the sheep-walks in ecological character.

By 1900, our upland bird fauna was partly original and partly influenced by these land use practices. Some species, especially the predators, had undergone local decline or extinction, but some birds favoured by these land uses occurred at high densities. Upland land use in more recent decades has tended to diversify and intensify, and has had an increasingly adverse effect on bird life. Since 1940, an estimated 30% of our uplands have been transformed or modified by land use practices, involving the loss of heather and bilberry to grass and bracken, and the loss of various types of open ground to conifer plantations and improved grassland or even arable (NCC, 1984; Woods and Cadbury, 1987). These changes, stemming from agriculture and forestry, are the most important in scale, but they are accompanied by others which have significant local impacts: water use, energy generation, mineral extraction, roads and buildings, defence use, recreation and game management. The effects of these activities on upland birds have been discussed under the different species, and are summarised below. The trends they have established mostly continue and some may even become enhanced. Beyond these are the more subtle but insidious effects of pollution in its various forms, including the especially pervasive atmospheric kinds.

Post-1900 changes in upland bird populations attributable to human activity

Agriculture

(a) Improvement of marginal land (drainage, ploughing, fertilising, re-seeding, herbiciding, increased stocking).
Decrease in Curlew, Snipe, Redshank, Lapwing.
Probable decrease in Black Grouse, Yellow Wagtail, Skylark, Meadow Pipit, Whinchat, Cuckoo, Chough.
Possible increase in Oystercatcher.

(b)   Reclamation of hill ground (same as (a), plus enclosure.
Decrease in Red Grouse, Golden Plover, Dunlin, Curlew, Snipe, Merlin, Hen Harrier, Short-eared Owl, Chough, Ring Ouzel, Twite, Stonechat, Whinchat, Skylark, Meadow Pipit, Cuckoo.
May locally provide increased feeding opportunities for some of these species, but causes general loss of nesting habitat.
Increase in Lapwing and perhaps Oystercatcher.

(c) Loss of heather and bilberry heath to grassland and bracken (through heavy grazing, usually with repeated burning).
Decrease in Red Grouse, Merlin, Hen Harrier, Golden Plover, Stonechat.
Possible decrease in Ring Ouzel, Twite, Short-eared Owl.

(d) Increased stocking density of sheep (increased availability of carrion).
Increase in Golden Eagle, Buzzard, Raven, Carrion and Hooded Crows.
Possible decrease in waders, including Dotterel.

(e) Improved sheep husbandry (reduction in mortality and carrion supply, through dietary supplements, winter feeding, treatment against disease, in-bye or indoor wintering and lambing).
Decrease in Raven and possibly other carrion feeders.

Afforestation

(a) The first 10–15 years after planting (increased luxuriance of moorland vegetation, young trees forming open scrub).
Early disappearance of Dunlin, Golden Plover, Lapwing, Raven (locally),

Wheatear. Increase in colonisation of Short-eared Owl, Kestrel, Hen Harrier (locally), Black Grouse, Stonechat, Whinchat, Wren, Grasshopper Warbler, Tree Pipit, Linnet, Redpoll and other woodland passerines.

(b)   From 15 years until felling (closure to thicket forest)
Total disappearance of all ground-nesting birds of open submontane moorland, causing serious national decrease in Golden Plover, Dunlin, Curlew, Snipe, Greenshank, Merlin, Red Grouse, Ring Ouzel, Wheatear. Upland species favoured by stage (*a*) disappear again: Short-eared Owl, Hen Harrier, Black Grouse, Stonechat, Whinchat. Development of the bird community depends on whether thinning of the maturing forest occurs, but in general, there is a replacement of an open hill bird community by one of woodlands, with Song Thrush, Blackbird, Chaffinch, Robin, Great Tit, Coal Tit, Goldcrest, Dunnock, Tawny Owl, Sparrowhawk, Jay, Woodcock, Wood Pigeon. Of the rarer species, Crossbill and Siskin often become abundant, Long-eared Owls appear in small numbers, and very locally Redwings and Goshawk colonise.

(c) Clear-felling and ground vegetation re-growth (second rotation open phase, until thicket forest closes again).
Re-appearance in small numbers of some open ground birds: Red Grouse, Black Grouse, Short-eared Owl, Hen Harrier (occasionally), Curlew.

Fig. 67. Fleet Forest and Grobdale, Galloway. Blanket afforestation in the Southern Uplands, causing large-scale replacement of moorland birds by woodland birds. Little plantable land is left unplanted, and Golden Plover have declined even on the adjoining unplanted ground. Streams and lakes become acidified.

Abundance of open woodland, glade and scrub species: Willow Warbler, Robin, Tree Pipit, Yellowhammer, Grasshopper Warbler, Redpoll, Reed Bunting, Whitethroat, Whinchat. Important habitat for Nightjar in some southern hill areas (e.g. North York Moors).

Overall effects depend on the balance, within any one forest, between stages (*a*) and (*c*), in both absolute and proportional areas, and also on the extent of ground left unplanted, both within and outside the forest. Blanket afforestation over large areas causes population declines in Raven, Buzzard, Golden Eagle and Merlin, but these species may keep going, and even breed within forest, in areas where the forest blocks are sufficiently dissected, leaving enough open ground overall.

(d) Effects outside the forest

Where afforestation substantially reduces the hunting range of the larger predators (Raven, Buzzard, Golden Eagle, Merlin) pairs will drop out, even though substantial areas of open ground remain. Moor burning usually ceases on ground adjoining plantations and the resulting increased luxuriance of moorland vegetation is believed to have caused local decline in Golden Plover and Dunlin. The possibilities for increased predation on eggs and young of moorland birds adjoining new forests are being studied. In parts of Wales and Galloway, Dippers have declined evidently through enhanced stream water acidification within afforested catchments.

Fig. 68. Kielder Reservoir and Forest, Northumberland. The moorlands of the North Tyne valley have been extensively covered with conifers: enclosed grasslands and meadows of the valley floor have lately been submerged by creation of the huge lake. Remnants of rush-grown marginal land with sheep still represent the once predominant land-use of the area.

Table 5. *Impact of afforestation on moorland birds*
Over 1 million ha of new conifer forest are in the uplands and so have reduced upland bird populations on a large scale, but little quantitative information is available on the precise effects.

Flow Country (Sutherland & Caithness)
From surveys of 77 sample areas, Stroud *et al.* (1987) calculated average wader breeding densities according to three types of habitat identifiable from maps. By measuring the areas of these habitats for the whole Flow Country and in the 67 000 ha already afforested or destined to be planted, they estimated losses as 912 of 4900 pairs (19%) of Golden Plover, 791 of 4620 pairs (17%) of Dunlin, and 130 of 760 pairs (17%) of Greenshank. Losses for other species have not been estimated, but most of those mentioned in Chapter 6 have been affected. One of the very few mainland nesting places of Red-necked Phalarope has been degraded and deserted by the bird.

Southern Cheviots (Northumberland, Cumbria, Roxburgh)
58 000 ha of afforested moorland between Redesdale, Bonchester Bridge and Bewcastle. Graham (1934) recorded that during spring on the Bewcastle – Gilsland sector 'Curlews were present in large numbers throughout the whole area'; that Golden Plover 'were comparatively abundant . . . at a pair to each quarter of a mile of moorland crossed'; and that Dunlin were sparingly but widely distributed. Dr H.M.S. Blair stated that the North Tyne moorlands were especially good ground for these three species and Merlins, and the area also had good grouse moors (Chapman, 1907). Allowing for habitat variations and applying conservative estimates of breeding densities, the following are estimated as minimum losses: Curlew, 1750 pairs (3 pairs/km^2); Golden Plover, 1200 pairs (2 pairs/km^2) Dunlin, 200 pairs (>1 pair km^2 of flow); Red Grouse, 11 600 pairs (20 pairs/km^2); Merlin, 25 pairs (5 pairs/100 km^2). Across the whole of the Cheviots, loss of 8 out of 20–25 pairs of Ravens is attributed to afforestation (Marquiss *et al.*, 1978).
Information for other regions is extremely patchy.

Southern Uplands
Curlew: a huge area planted, often where nesting density was high; at least 5000 pairs lost.
Golden Plover: less numerous on planted areas than in the previous district, but probably at least 50% of Galloway population lost.
Dunlin: probably 150–200 pairs lost from colonies on flows of Ettrick–Teviot–Esk watershed, central Kirkcudbrightshire, Wigtownshire and South Ayrshire; representing a substantial proportion of total Southern Upland population.
Snipe: a huge loss of suitable habitat, but no estimate of numbers available.
Wigeon: decline of isolated population on afforested lochs in Ettrick.
Raven: of 100+ regular pairs (1946–60), at least 31 pairs believed lost through afforestation (Marquiss *et al.*, 1978; Mearns, 1983).
Buzzard: stable Galloway hill population of at least 25 pairs declined to 1–2 pairs in 1986 (p. 49).
Merlin: of at least 8 pairs known in Cairnsmore of Fleet area in mid-1970s, only 2–3 pairs left by 1988 (A.D. Watson).
Golden Eagle: in the small Galloway–Carrick population of 4 pairs, Marquiss *et al.* (1985) connected low breeding success in 3 territories, and eventual loss of one of these territories, with advancing afforestation. A second pair has since dropped out.

Table 5. (*continued*)

Eastern Highlands
At least 88 000 ha of grouse moor planted in Stirling, Perth, Angus, Kincardine, Aberdeen, Banff, Moray, Nairn and Inverness. Good habitat for Red Grouse, Golden Plover, Curlew, Hen Harrier, Merlin and Short-eared Owl.
On Speyside, breeding grounds of 7–10 pairs of Greenshanks now afforested (D. Nethersole-Thompson).

Western Highlands
Watson *et al.* (1987) predict, on the basis of decrease already observed, that the Golden Eagle breeding population in the south-west Highlands, south of Oban, will decline to about 50% of 1960 levels by 1995–2000; this will represent a loss of 20 pairs to afforestation *already* carried out. This study also envisaged that the present Golden Eagle population of the Highlands (*c.* 420 pairs) could decline by at least 20% as a result of existing afforestation by the year 2015.

North Wales
At least 12 000 ha of the best Red Grouse, Golden Plover, Dunlin, Merlin and Short-eared Owl ground has been planted, especially on the Denbigh Moors, Berwyn Mountains and Llanbrynmair Moors.

Water supply and hydro-electric power (raising natural lake levels or creation of new reservoirs; unnatural shore-lines, usually with frequent and excessive fluctuation in water levels; often tapping of inflow streams and reduction in outflows)

Effects include decline (through flooding of breeding habitats) of valley bottom breeders such as Lapwing, Snipe and Redshank. Probable reduction in habitat feeding value for Greenshank, Common Sandpiper, Dipper. Species such as Black-throated Diver variably affected, according to whether islands are created or submerged. Some waterfowl attracted anew, whether as breeders (Wigeon, Common Scoter, gulls) or wintering birds (various ducks and other waterfowl).

Mineral extraction and quarrying

There is local disturbance when workings are active, but the scale of impact is usually not important. Abandoned workings become important breeding places for Raven, Chough, Peregrine, Kestrel, Stock Dove, Ring Ouzel, Wren and Jackdaw. Mining and quarrying have probably boosted populations of the first three species in areas where other natural nesting places were 'saturated' or not used (Chough).

Defence use

This may be beneficial through preventing other forms of disturbance, but may itself create damaging disturbance where use is heavy. Most defence areas happen not to be particularly important for upland birds.

Recreational use

(a) Hiking
Disturbance by hill walkers is mostly extremely local in adverse effects on nesting birds. Picozzi (1971) found no difference in Red Grouse breeding performance and shooting bags on Peak District moors with unrestricted public access for hiking, compared with those where access was limited to rights of way. Golden Plover are frequently kept off eggs and chicks by hikers in popular areas of the Peak District; birds with chicks usually alarm at intruders within 200 m, but effect on breeding performance is more serious where visitors allow dogs to run loose. Nesting may cease where disturbance is severe (Yalden and Yalden, 1988). In the Peak District, W. Gibbs (personal communication) believed that a high density of Golden Plover during 1928–39 on Big Moor, Totley Moor and Burbage Moor declined first through intensive military training during 1940–45, and failed to recover when the area became an especially popular recreation area. Effects are more noticeable where public roads allow easy access to moorland by vehicles, and breeding density of Red Grouse, Golden Plover, Curlew and Lapwing may decrease near where cars often park. On high ground, Golden Plover, Dunlin, Dotterel and Ptarmigan often nest close to popular hiking tracks, and frequently hatch off. In Lakeland, 6 Merlin tree nests (all successful) were within 20–100 m of a popular walking track, and 2 of them were within 300 m of a much-used picnic site. Open public access increases the risk of uncontrollable and destructive moorland fires. Several tor-nesting sites of Ravens on Dartmoor have been deserted through hiking disturbance, but the birds have probably found alternative tree sites on the edge of the moor. Peregrines may be inhibited from using tor nest sites by such disturbance.

(b) Rock-climbing
The ability of crag-nesting birds to adapt to climbing depends both on its frequency and the size of the cliffs. Several small gritstone crags in Northumberland have been deserted by Ravens and Peregrines since they became popular climbing grounds, after 1950. Elsewhere, climbing frequency has evidently detered these species from occupying the low (20 m) m) but suitable whinsill crags along the Roman Wall, and the gritstone scars in the southern Pennines. In Snowdonia and Lakeland, Peregrines have either moved to smaller and less climbed alternative crags or have adapted to climbing. Though some nests continue to fail through disturbance, some individual falcons have become remarkably bold, continuing to

incubate or returning quickly to their nests when climbers are at work nearby. Ravens have also adapted to climbing on some cliffs. Golden Eagles are shy birds and sensitive to climbing disturbance, but in the popular climbing areas of the Highlands, quieter alternative cliffs are usually available.

(c) Fishing

On many streams and lakes, riparian and aquatic birds have to endure much disturbance by fishermen during the nesting season. There is evidence of local decline in the Common Sandpiper, and the Oystercatcher, Ringed Plover, Dipper and Grey Wagtail are vulnerable to adverse effects on breeding. River and lake edge feeders such as Greenshank and Dunlin, and lake swamp nesters such as Black-headed and Common Gulls, may be discouraged by frequent intrusion, and most upland waterfowl whose

Fig. 69. Fellow creatures of the crags: Peregrine and climbing disturbance.

young take to the water when small are vulnerable to boat and shore fishing. Islands otherwise secure are often visited by fishermen, and the poor breeding of both Black-throated and Red-throated Divers in some areas may be partly the result of fishing disturbance. Keeping incubating birds off their nests often increases predation risks to the eggs. Altogether, at least 20 species are exposed to this effect, but there is little information to show its significance.

(d) Downhill skiing

This is a major post-1945 development, mainly in the Highlands. Its direct effects are limited: decline in Ptarmigan in the main skiing Cairngorm corries has occurred through the high risks to birds hitting ski-tow cables. Although it is a winter sport, a crucial effect has been to open up hitherto remote high ground for spring and summer access by walkers. The vehicular road and chairlift facilities carry large numbers of visitors to within easy reach of the Cairngorm–Ben Macdhui plateau, which has seen a huge increase in human pressure. There is no evidence yet of adverse effects on the breeding birds here, but one result has been to attract crows and gulls, which scavenge discarded food and then search for nests. Snow Bunting parties similarly attracted in winter are a bonus.

(e) Other

One of the more worrying changes is intrusion by new forms of transport. A notable feature of the past 20 years is the vastly increased penetration of the hill country by unsurfaced roads, bulldozed into remote uplands to simplify access for shooters, keepers and shepherds. Though mostly private, they also ease access for walkers. The new forests are also served by systems of well-made roads. Many once rough public roads across the hills have also been metalled, thereby increasing numbers of picnickers and walkers. A potentially more serious factor is the increasing use of 'all-terrain vehicles', which do not need roads and which bring motorised disturbance to high ground with nesting Golden Plover, Dunlin, Dotterel and Ptarmigan. Hang-gliding has caused failure of several Peregrine and Raven broods, and is believed to be responsible for the decline of one especially dense Pennine population of Golden Plover and Lapwing. Micro-light aircraft bring the scope for much increased disturbance to once remote places. Boating on upland lakes is another increased activity, and may have contributed to the virtual desertion of Loch Morlich as a Greenshank feeding haunt. Finally, there is the disturbance that well-intentioned bird watchers may cause.

Game management

Grouse management has maintained large areas of heather-dominated moorland, by controlled burning and avoidance of high sheep stocking levels. This has benefited Golden Plover, Merlin, Hen Harrier, Stonechat and perhaps other species (p. 106). Control of Foxes and Crows may enhance breeding performance of other birds besides Grouse, and may help to maintain dense populations of certain waders. Decline in Golden Plover has followed abandonment of grouse moor management, and may have involved increase in nest predation. The direct effects of predator control have, however, had important negative impacts on certain rarer birds. Leslie Brown (1976) has dwelt at length on the problems, but the present situation may be summarised as follows.

Raven. Largely absent from keepered grouse moors, and has declined in the Pennines and Spey valley during the past 20 years, evidently from this cause.

Buzzard. Few breed on grouse moors, but many are killed there, dispersing from western strongholds. Almost absent from the Yorkshire Pennines at present.

Golden Eagle. Few breed on grouse moors, but many are killed on those of the eastern Highlands. Failure to recolonise the Pennines probably attributable to keeper destruction.

Red Kite. Slowness of recovery may be partly the result of high mortality in both lowland and upland game preserves.

Hen Harrier. Grouse moor a major habitat and continual attempts at establishment there, but almost universally detested by grouse men, and subject to great persecution, all of it illegal.

Peregrine. Formerly given similar treatment to Hen Harrier, and widely and ruthlessly destroyed. Regarded with more relaxed attitudes since the post-1956 crash and even protected on some grouse moors. Signs of a return to former persecution locally during the past few years.

Merlin and Kestrel. On the whole nowadays tolerated on grouse moors, and even protected on some; but a small number of keepers remain unpersuaded of their insignificance as game predators, and continue ruthless destruction.

Goosander and Red-breasted Merganser. Generally disliked by fishermen and killed in quite large numbers (including under licence) in Scotland.

Gun and gintrap are still used quite deliberately against these predators by some keepers, but probably the biggest problem for the habitual carrion feeders (Raven, Buzzard, Golden Eagle, Red Kite) is the illegal use of poisons, especially in meat baits. These are widely put out, and by farmers and shepherds as well as keepers, to kill Foxes and Crows. Egg baits are less used nowadays, since they are so obvious to those who might inform the police, but meat baits are difficult to detect and are often used with impunity.

Nest robbing

Illegal collecting of bird skins or mounted showcase trophies appears to be a minor issue nowadays, but the RSPB believe that there are still several hundred illicit eggshell collectors and a similar number of others who take eggs and young of raptors for falconry (or, more often, 'hawk keeping'). The effects of the great age of egg-collecting (roughly 1880–1940) are difficult to assess, but clearly placed a severe drain on the breeding success of many rare and local species. Egging probably contributed to the final demise of the Osprey and Sea Eagle and to the near-extinction of the Red Kite. Its worst effects were probably on rarer species already under severe pressure from other causes. Birds such as the Peregrine appeared to hold their numbers remarkably well in the face of relentless collecting in certain districts. The reduced recruitment nevertheless probably prevented them from increasing and achieving a wider distribution. The recent recovery of the Peregrine to unprecedently high levels in certain districts appears partly to reflect a high output of young which could not have occurred if egg-collecting had continued at its pre-1940 level. The fact that it did not so continue is a tribute to the combined effort on protection and education in recent years.

The present consensus is that egg-collecting is an outmoded and undesirable activity, which has a considerable nuisance effect and may threaten the conservation of some birds. There is legal protection for the eggs of nearly all species. Falconry is still treated as a legitimate elite sport, but a cloud is cast by the continuing scale of illegal taking of young raptors and eggs to hatch, and the use of captive breeding as a 'cover' for this. There is understandable reluctance to licence taking from the wild until the considerable traffic in illicitly taken raptors can be eliminated. It is an activity which could rapidly increase if protection efforts were relaxed.

Genetic 'finger-printing' by DNA analysis now allows a check on the parentage of birds allegedly captive-bred. It will thereby help to close another loophole in illegal taking from the wild.

Pollution

The episode of widespread poisoning and decline of birds of prey caused by the persistent organochlorine insecticides of agriculture appears, happily, to have passed, though low-level contamination by these residues remains widespread. The Peregrine, affected in the uplands by the use of these chemicals mainly in the arable lowlands, has made a spectacular recovery. The depression in breeding success found in Golden Eagles and Buzzards in certain upland districts also disappeared rapidly when the organochlorine insecticides were withdrawn from use in sheep dips. Although Newton and Haas (1988) have found no conclusive evidence that the serious local decline in the Merlin is caused by its continuing contamination by organochlorine residues (including PCBs) and mercury, the levels of these substances found in the population remain worrying.

More concern now focuses on the fall-out of acidity from atmospheric pollution, mainly by sulphur dioxide and nitrogen oxides. Acid deposition in the uplands is intensified by the heavy rainfall, and is subject to further enhancement in heavily afforested catchments through the filtering effect of coniferous trees. This acidity is transferred especially to lake and stream waters, where it locally had adverse effects on Dippers (see p. 82), and in some lakes has reduced or eliminated the larger fish on which other birds feed. Evidence for the increased acidification of soils and peats is only now beginning to accumulate, but such an effect has many implications for carrying capacity of uplands for birds and other vertebrates which will have to be examined in future.

The latest pollution concern could prove to be the most important of all in its global effects. Scientists have lately reached a virtual consensus that the World *will* experience a 'greenhouse effect' through increasing carbon dioxide and other gaseous pollution. The resulting increase in global temperatures will cause a widespread rise in sea level, with predictable consequences to low-lying coastal lands. There are wide differences in prediction of the rate and magnitude of the changes and their regional manifestations. It is conceivable that Britain and Ireland could become colder instead of warmer, and rainfall might either increase or decrease in different regions. Some of the more extreme scenarios have alarming portents for our wildlife as a whole. Even a more modest average warming of 2–3 °C in annual temperatures over the next 50–100 years in Britain would rapidly restore conditions approximating to those of the post-glacial climatic optimum. There would be scope for trees to grow at much higher elevations and for montane habitats and wildlife to become correspondingly reduced. Fringe species of upland birds might move upwards and retreat still further from lowland areas. Retreat of polar ice, northwards advance of the taiga forests, and rising sea level might have profound effects on the distribution of many birds of the Arctic regions and could alter migration patterns. Local return to colder conditions would, predictably, reverse some of these tendencies. The prospects are uncertain but awesome.

Summary

In the sheer geographical scale of their impact in reducing or transforming the upland bird fauna, agriculture and afforestation far outweigh all the other land use activities. Game preserving continues to have a widespread effect in holding down the range and numbers of certain predatory birds to far below their potential.

Finally, although their impact on the birds themselves may be negligible, some developments have a shattering effect on the wilderness qualities of the uplands, and thus detract greatly from the enjoyment of the many people who wish to see birds in unspoiled settings. These developments include tree-farming afforestation; organised skiing facilities; dams, pipelines and unsightly reservoir draw-down zones; moor-gripping; new hill roads and vehicles; pylons and powerlines; and hill-top buildings such as radar installations.

Conservation aims and measures

The uplands and undeveloped coasts are the only extensive areas of natural and semi-natural habitat now remaining in Britain and Ireland, and their wildlife is a tremendous national asset to both countries. The upland bird fauna also has international importance. The scale of past losses, the fragile present status of many species, and the portents for continuing depletion, thus point to the need for a vigorous strategy for the conservation of our upland birds, but as part of an overall approach dealing with other wildlife. The conservation of habitat (i.e. physical environment and vegetation) is fundamental to the needs of any species, but more direct methods of management and protection are necessary to conserve many kinds of birds. I believe that bird conservation in each region of Britain should aim to do the following.

1. Maintain or enhance the species diversity of bird assemblages characteristic of natural and semi-natural habitats. Where increase in diversity can occur by removing human impediments to the natural spread of species, this should be done if practicable. Deliberate reintroduction of lost species may be allowable but each case has to be considered on its own merits. Large-scale creation of artificial diversity, by replacing natural or semi-natural habitats with wholly man-made types (e.g. coniferous plantations on moorland) is not an acceptable aim.
2. Maintain, within the limits of normal fluctuation, populations and breeding densities of all species protected during the breeding season. When species show long-term decline below a desirable level, remedial

measures may be necessary, but based on a clear understanding of the causes, which may require research.

3. Protect and encourage Schedule 1 species (22 upland species), especially those rare and endangered. These may be stable, yet vulnerable to habitat loss or other disturbance. There will always be species that maintain a mere foothold in Britain, if only for reasons of marginal climate, but they should be given the best chances of survival here.

4. Ensure that the case for control of Schedule 2 Part II species (those which may be killed or taken by authorised persons at all times), and for issue of any licences to kill otherwise protected species, is based on well-substantiated evidence, and achieved by legal methods.

Above all, there is a need to minimise further loss and damage to upland bird habitats through adverse human impact, and to ensure that land management methods are, as far as possible, benign. The measures available to meet these objectives fall under four headings.

Protection of important areas

The cornerstone of conservation strategy involves setting aside the best areas for birds as nature reserves or sanctuaries, to protect and manage the species concerned as the primary objective of land use. This is the strongest of the available measures, but has its limitations, as we shall see.

The 'best' areas for upland birds are those which have the greatest diversity of species overall, the densest populations of dispersed species, the largest populations of colonial species, or the greatest variety and numbers of rare species. Reserve and Site of Special Scientific Interest (SSSI) selection have generally aimed to represent adequately the total range of variation in ornithological habitats, but it has increasingly been felt that all areas above a certain level of quality should be thus protected. Diversity in birds tends to be related to variety in habitats, so that ecologically varied uplands are often the best. Chapters 2–7 have, however, shown the limits to such variety within any one massif. Selection of upland areas for reserves and SSSIs should thus try to represent the range of variation on a local, regional and national scale. The inclusion of fringe populations of species with restricted distribution is important.

While they are the most effective conservation measure, the protected areas can achieve far less than is needed in total. The upland SSSIs cover about 633 500 ha, of which 87 260 ha are held by the NCC, mostly as National Nature Reserves. Of these 87 260 ha, only 36 300 ha are owned, compared with the less satisfactory lease or reserve agreement tenure; and current Government policy favours the last two arrangements. The RSPB have 8920 ha of upland reserves, and the Wildlife Trusts within the umbrella of the Royal Society for Nature Conservation own another 10 500 ha. The two National Trusts own quite a large area of upland, which is protected in some degree, though not exempt from farming and forestry modification. Moreover, many of these other areas fall within the SSSI total. SSSI status is no guarantee against damaging activities, and their long-term effectiveness will depend on continuing Government willingness to finance compensation arrangements for agricultural or forestry development profit forgone, or to use compulsory purchase when attempts to reach such settlements fail. The Wildlife and Countryside Act 1981 failed to give SSSIs increased protection against developments which fall within planning law, and even when some of these are prevented, there is open-ended scope for new and often different proposals in due course.

The various categories of protected area total around 750 000 ha or 11.5% of the remaining extent of uplands. They contain some of the particularly good areas for upland birds, holding the bulk of the populations of rare or local breeders such as Dotterel, Snow Bunting, Red-necked Phalarope, Wood Sandpiper, Purple Sandpiper, Great Skua and Arctic Skua. Yet, few upland birds show any marked degree of colonialism, and many are highly dispersed, with a wide distribution, so that the bulk of their populations fall outside these protected areas. These species are the biggest conservation problem. Even if further proposed nature reserves are established, their total contribution to the conservation of upland birds will remain relatively small.

The 9 upland National Parks of England and Wales cover 705 605 ha of uncultivated mountain and moorland (Countryside Commission, unpublished), though this figure contains a substantial extent of the protected areas already described. Those in the Brecon Beacons, Snowdonia and Lakeland support important populations of Peregrines, Ravens and Buzzards; and probably most of the inland nesting Choughs in Britain are in the Snowdonia Park. The Yorkshire Dales, North York Moors and Peak District National Parks contain good numbers of Red Grouse, Curlew, Golden Plover, Dunlin and Merlin (not the Peak nowadays). These six Parks together give good representation of common or widespread species such as Meadow Pipit, Skylark, Wheatear, Ring Ouzel, Dipper, Grey Wagtail, Cuckoo, Carrion Crow and Common Sandpiper. The Dartmoor and Exmoor Parks have good Raven and Buzzard numbers, and fringe populations of several species, but have less diversity and overall importance than the others. The Northumberland Park excludes some areas now extensively afforested, which were once the best bird ground in the Cheviots. The Areas of Outstanding Natural Beauty (AONBs) include the Northern Pennines and Bowland Fells, two particularly important areas for moorland birds; but of the others, only the Shropshire Hills are significant for upland species.

How effective are these wider designations in protecting birds, anyway? The National Parks have limited control over agricultural improvements damaging to wildlife, and still less control over afforestation. The AONBs are, if anything, still weaker in both respects. Both the Parks and AONBs must, on the whole, be regarded as better than nothing, and we must hope that they become more effective in the future. In both, recreational use takes precedence over nature conservation, but any conflict between the two is usually of only minor importance to upland birds. Scotland has no National Parks, and its National Scenic Areas are a feeble substitute which seem unlikely to confer any meaningful benefits on wildlife. Even many Third World countries have National Parks, but proposals to establish them in Scotland were voted out in 1948 by the combined power of landowning and local political interests, which have also successfully squashed any subsequent attempts to resurrect definite proposals.

This leaves a large proportion of our uplands without any protection against developments inimical to wildlife, and dependent on the uncertain balance between the intentions and attitudes of the owners, occupiers and their employees plus outside development interests; as compared with the wishes of those bodies and individuals whose interests and activities impinge on the uplands, but who have no rights of ownership or occupation. The pressures of the various developments and the changes they have wrought in the uplands, are an expression of the different 'rights' claimed by separate interest groups. Conservation of birds and other wildlife represents only one of these interests, though it forms a natural alliance with that concerned with the preservation of scenic beauty, and the use of the uplands for quiet recreation. The due share of conservation in the wider setting of the uplands as a whole is a socioeconomic and political issue, and this leads into discussion of the second of the conservation measures.

The wider countryside approach

This has to sit within rural land use policy, and so depends on conveying the strength and validity of the conservation concern, but also on devising workable methods for harmonising this with the interests of the resident community. It depends on informing and persuading society at large to cherish our wildlife as a key part of the natural heritage, through public relations and educational activities. It also involves working out management objectives and prescriptions, and ways for supporting these through the financial mechanisms available.

Public opinion could do a good deal, if it were properly marshalled. At bottom, much of the continuing damage and loss to upland habitats and wildlife results from the activities of a relatively small number of people whose rights of land ownership and occupancy are mostly heavily subsidised from public funds. Their concern to protect or improve their financial position and livelihood is natural, but has to be set against the wishes of the much larger (and still increasing) number of people to whom the uplands are a vital recreational and spiritual resource, and who contribute significantly to the Treasury funds which shore up the residents' position. The case for further hill land reclamation is economically absurd, yet it will go on in many a desperate attempt for survival, until there is a sensible alternative. Hardly any of the public subsidies involved are recouped by some economic return to the nation: mostly they represent money written off as a social subsidy to maintain small human populations in continued occupation of the hill country.

Probably most people who live elsewhere and visit the uplands in their spare time would be pleased to see the rural communities remain in their present occupation of these areas, and prepared to make their own contribution in support through Government subsidies. But such attitudes will be more likely to prevail if the hill occupiers are prepared to manage their land in ways which maintain or, better, enhance those features which the visitors wish to see and enjoy. The principle is already accepted as the basis for topping-up farmers' incomes within Environmentally Sensitive Areas, though this is at present a very limited and localised voluntary scheme whose results have yet to be measured and assessed. Nevertheless, it is seen as a possibly much more widespread solution to the problems of agricultural over-production, uneconomic hill farming and the drift of people from the hills and to the towns. With the steady tipping of the economic scales against hill farming, and the probability of a forthcoming sheepmeat surplus, this must be the least of several evils for many owners and occupiers.

Since 1920 the combined efforts of the Forestry Commission and private forestry have established 1.2 million hectares of new forest, mostly of conifers and in the uplands. The resulting doubling of the depleted forest cover of Britain from 5% to 10% would seem a commendable achievement, in principle. The initial results for birds are often impressive, but within 15 years, the moorland bird fauna is exchanged for one of woodlands. Probably most naturalists would have accepted the case for some of this afforestation, or in certain areas, and subject to conditions, so that intrusion into important scenic and wildlife areas was avoided. But successive Governments have yielded to the pressure of the insatiable forestry lobby in endorsing an open-ended 'policy' of afforestation of hill ground. The limits appear to be set only by the extent of plantable land (a somewhat elastic concept) and in some districts there are worries whether any significant areas of submontane upland will eventually remain.

If this seems an exaggeration, go and look at the evidence: on the moorlands of Kielder in Northumberland; Craik, Eskdalemuir and Galloway in the Southern Uplands; and Knapdale–Kintyre in Argyll. The remote Cheviot headwaters of the North Tyne, Irthing and Lyne once had some of the most beautiful moorlands in the country, and were the nearest to real wilderness that England could show. They are now extensively blanketed with conifers, and still the planting goes on. In Scotland, a private forestry spokesman has provided the following scenario: 'In the year 2026, I do not foresee blocks of forestry on a moorland landscape, but the stark moorlands of SSSIs and common grazings isolated in a forest scene' (Ogilvy, 1986).

Matters came to a head over the great flows of east Sutherland and Caithness. For many years, forestry in this far northern region of Scotland was only patchy and mainly around the drier edges of the peat moorlands. The foresters were, as elsewhere, unforthcoming about their intentions in the district. Noting the draining and planting of wet bogs farther south, and the spread of forests around Lairg, the NCC and RSPB began systematic surveys in the late 1970s to identify the most important wildlife areas. They were too late. By 1980, the Forestry Commission and the private company Fountain Forestry were rapidly acquiring large areas of flow ground, and pressing on with planting programmes. By 1988, 67 000 ha of peatland in this region were either afforested or programmed for planting. This is 16% of the total blanket bog area here and losses have fallen disproportionately heavily on some of the best areas for birds, especially the massive flow ground around Aultnabreac. They have also been scattered piecemeal to affect 33 out of the 41 major catchments.

The afforestation of the Flow Country has caused particular alarm, not only because it holds the most diverse and unusual moorland bird community in Britain, but also because the habitat, blanket bog, is one of the rarest vegetation formations in the World, and this may be the largest single expanse which exists. The Government appears to accept that no more than half the total Flow area of 400 000 ha deserves conservation safeguard, and is disposed to allow the rest to be subject to further planting to a total of at least 100 000 ha according to the vagaries of the land market.

If all this afforestation were of immense benefit to the nation, the state support for it would be more understandable. Two independent Government enquiries, by the Treasury in 1972 and the National Audit Office in 1986, have exposed the unsoundness of the economic and social justifications pleaded for continuing upland planting. Virtually the whole of the new forests have been paid for largely out of the public purse. State forestry has been run at a substantial loss ever since its inception, while private forestry became profitable only because heavily subsidised investments were convertible into tax-free capital gains. While tax relief for private forestry has since March 1988 been replaced by direct grant aid, the level of taxpayer support is only marginally less than before. And the 'downstream' part of the forestry industry which uses home-grown timber is also supported by a substantial, though usually undisclosed, level of Government grant-aid to construction of processing plant.

It is high time that state subsidy to afforestation, whether by the Forestry Commission or private sector, was restricted to planting which, while complying with minimum environmental standards, can reliably meet the economic rate of return expected from normal public sector investment. This would eliminate a great deal of further planting of high or wet hill ground where timber production is poor or thinnable crops unlikely. The arguments for job creation should also be required to demonstrate greater cost-effectiveness than investment in alternative enterprises in the same region. Forestry is now required to achieve a 'reasonable balance' with other interests in any area, but in the continuing absence of Government guidance which adequately meets the needs of other interests, this singularly empty phrase tends to signal business as usual. Over the whole country, some 270 000 ha, almost entirely in the uplands, were already approved for planting in March 1988 and may yet be afforested with tax relief under the transitional arrangements.

There is a good case for restoring *some* semi-natural woodland, of native trees, in many deforested uplands, but in sympathy with the needs of open hill wildlife and landscape. The new and artificial conifer plantations are not an acceptable substitute either for original woodland or for the semi-natural hill ground derived from this. While they bring a limited woodland bird fauna, including a few rare species, to many largely treeless uplands, this is not an argument for allowing their further expansion. This fauna is now sufficiently well represented in existing forests and we are bound to regard its further replacement of upland bird populations as undesirable. The 'tree farms', with their thickets of exotic trees, are a particularly unacceptable replacement for habitats, such as peat bogs, which have been naturally treeless for thousands of years. Afforestation has caused a greater loss of upland birds than any other factor during this century and, unless there are radical changes in policy, the species which have already declined can be expected to show still more serious losses, especially in Scotland.

All major new afforestation schemes (above 25 ha) should be required to sit within a regional land use strategy and be subject to environmental impact assessment. Only then can there be an adequate mechanism for ensuring 'reasonable balance' with other interests, and avoiding further unacceptable damage to wildlife interests, scenic beauty and other environmental concerns. There must also be a predetermined ceiling on the final extent of afforestation in any district. The Forestry Commission has proved far too partial a body to be entrusted with the administration of such an approach, and there would seem little option other than to place this under the existing process of planning control of development, as an appropriately stringent mechanism of adjudication between conflicting interests. The most helpful recent development is Government support for the positive idea of encouraging afforestation on the over-producing arable lowlands, where it can produce good timber and become an environmental benefit. This must, however, be regarded as an *exchange* for planting the uplands, and not as an addition, as the forestry lobby would so obviously prefer.

Management as deer forest and grouse moor have saved large areas of the uplands from more damaging modifications, yet they, too, are increasingly under threat of afforestation as the economics of these sporting interests wane. While grouse moors are certainly good for some birds, especially waders, and their heathery expanses are usually scenically attractive, many of them are less interesting than they might be because of the poor representation of predatory birds, largely as a result of persecution. Many ornithologists would gladly exchange the artificially high grouse numbers and model muirburn for a greater assortment of predators and scavengers and more casually managed moorland. Even if the price was also a reduction in density of birds such as Golden Plover, it would be worth bearing for the addition of Raven, Peregrine, Hen Harrier and Golden Eagle living in sanctuary on the moors. Surely, it is not too much to ask that there should be a few areas of heather moorland where Nature is allowed to evolve a balance within its bird community closer to that existing before the advent of Man. This much ought to be achievable on National Nature Reserves, yet even here it seems now that the pursuit of field sports – with all that this entails – is to be a principal objective. Britain's National Parks have long been a derision by World standards, and if we are not careful our National Nature Reserves will be held up to ridicule, too.

It is worrying, also, that interest in maintaining Grouse stocks – on which the viability of the moors is claimed to depend – is causing some people to think that an answer lies in the removal of protection from certain birds of prey. Ornithologists have welcomed the slow spread of enlightenment among game preservers in regard to the protection of some birds of prey, and any attempt to put the clock back on this issue would be greeted with uproar. The present scale of illegal killing is still large enough to restrict the distribution and numbers of several species, and the only acceptable change would be further increase in respect for the existing law.

Other land use activities will continue to chip away at the uplands. Mining and quarrying are, from past experience, not too worrying, provided they avoid important physical or biological features and sensitive scenic areas, and are not on too large a scale. Water supply in north-east England is believed to be over-provided, through the Kielder and Cow Green Reservoirs, but there may well be pressure to create other reservoirs in Scotland and Wales, and there has been a revival of interest in hydro-electric schemes in the Highlands. Two new proposals for hydro projects – incredibly enough, within the Wester Ross National Scenic Area – have been withdrawn for the time being. There are vested interests here who would happily see every glen in Scotland scarred by the works of hydro engineers. The effects on birds may not be great, but for those who enjoy seeing birds in unspoiled settings, they are a disaster. Further schemes are of extremely doubtful economic validity, and quite enough damage has already been done by the existing ones.

Other intrusive developments, such as radar masts, windmills, power lines, and hill roads for various purposes, will doubtless continue to threaten the peace and beauty of unspoiled uplands. Any proposals for such constructions should require very convincing justification and consideration of alternative sites, before being accepted. The availability of public subsidy, including tax relief, for the making or renewal of hill roads should be subject to especially strict scrutiny. The biggest of these intrusive development threats comes from the pressure for yet more opening up of new hills for winter sports, notably in the form of facilities for downhill skiing. Where such extension occurs there is also pressure to use chairlifts for conveying visitors to the high ground during summer, and so increasing access to the mountain tops. While the effects on breeding birds are as yet uncertain, there will assuredly be a further loss of wilderness character.

All these developments other than agriculture and afforestation have to be considered under the open and democratic processes of planning law. It is then up to the conservation side to generate effective lobbying in defence of its interests, including birds, so that this exerts the appropriate political weight at all levels from central Government downwards. The European Communities' requirement for member states to apply environmental impact assessments to major developments is potentially helpful, though its force in Britain has already been somewhat diluted, notably by restricting its application to afforestation to areas which are already covered by statutory conservation designations.

Pollution problems mostly originate far outside the hill country, and have to be tackled by administrative measures to reduce or eliminate environmental contamination at source. Constant vigilance is needed over new pesticides, both in prior screening through the Pesticides Safety Precautions Scheme and through field scrutiny to detect possible effects on birds. The possible problems with PCBs and mercury deserve closer study to measure their importance. There is general concern over chemical pollution of inland waters and the sea, and pressure to reduce this contamination.

Reduction in the atmospheric pollution producing acid deposition is another *cause célèbre*, but resistance by the British Government to achieving whole-hearted decrease in the main source, power station emission, is limiting progress here. Remedial work after the event such as the liming of acidified streams and lakes is not really an effective or satisfactory solution. Enhanced acidification by afforestation should dwindle if source emissions are reduced, but the run-off of fertiliser applied to boost tree growth is likely to remain as an inevitable side effect so long as trees are planted on poor upland substrates.

These aspects of pollution are eclipsed by the spectre of the impending global warming under the 'greenhouse effect'. Even without the more apocalyptic predictions, the possible consequences for humanity – let alone wildlife – are sufficiently alarming to require that every effort be made to obtain international cooperation in minimising emissions of the 'greenhouse' gases.

The bird protection laws

The Protection of Birds Act 1954 replaced earlier and disparate legislation with a more comprehensive and effective set of measures. These included a general presumption of legal protection for all species, except a defined short-list of those regarded as pests, and provision of special penalties in regard to killing and taking of a longer list of rare species or their eggs, or for wilful disturbance at the nest. Subsequent revision, including the latest, Part 1 of the Wildlife and Countryside Act 1981, have followed the same principle, with various improvements, including removal of loopholes.

The success of the bird protection laws in deterring the deliberate taking or killing of protected species and their eggs, and the disturbance of nesting rare birds, except under permitted circumstances, is always difficult to assess. In the case of upland birds, it does not prevent a good deal of continuing robbery of the eggs and young of many species, and the killing of many raptors. The law on wilful disturbance of Schedule 1 species is also widely flouted. Yet it has to be presumed that in the absence of this legislation, the situation would be vastly worse, with scope for a free-for-all onslaught which would threaten the status, if not the survival, of many birds. The particular combination of legal provisions, penalties and enforcement level does at least appear to be effective in containing deliberate, adverse human impact on bird life, to a degree which is satisfactory, except for certain species, mainly rarities and predators.

Heavier penalties and better enforcement might seem to answer the need for improvement, but the educational input to bird protection has probably been a major factor in the post-war improvements in people's behaviour, and this work needs much more effort. Bird protection law itself helps to generate a climate of feeling that birds must be important, and helps to raise public awareness of their value as part of our heritage of nature. Species protection legislation is thus an essential statutory adjunct to the other measures concerned especially with habitat conservation, and with public relations and educational work.

International conservation obligations

Wildlife conservation is a global concern, and Britain has recognised its international obligations through the signing of three relevant conventions. They are:

The Bern Convention on the Conservation of European Wildlife and Natural Habitats requiring the promotion of 'national policies for the conservation of wild flora, wild fauna and natural habitats, with particular attention to endangered and vulnerable species . . . and endangered habitats'.

The Ramsar Convention on Wetlands of International Importance, especially as waterfowl habitat, requiring contracting parties to promote the conservation of listed protected wetlands; identified according to eight criteria for assessment of international importance.

The World Heritage Convention requiring each State Party to nominate a list of 'cultural and natural properties' considered to be of 'outstanding universal value' against a set of carefully defined criteria. In addition, the EEC Directive on the Conservation of Wild Birds obliges Member States to give special attention to protection of the habitats of certain listed birds which are rare, vulnerable or otherwise needing particular attention; and requires that similar measures are applied to other regularly occurring migratory species.

There is a particular need to protect habitats and species which are rare on a global scale, and migratory species shared between different countries. The Bern Convention supports the protected area approach to upland conservation, and other general measures. The Ramsar Convention is relevant in the uplands especially to the conservation of blanket bogs, and the Caithness–Sutherland Flow Country is probably the British habitat best qualified for nomination to the World Heritage list. The EEC Birds Directive list includes: Black-throated Diver, Red-throated Diver, Red Kite, White-tailed Eagle, Hen Harrier, Golden Eagle, Peregrine, Merlin, Dotterel, Golden Plover, Ruff, Wood Sandpiper, Red-necked Phalarope, Snowy Owl, Short-eared Owl, Nightjar and Chough; and of the other migratory species, the Greenshank, Whimbrel, Dunlin, Temminck's Stint and Purple Sandpiper are particularly appropriate for special conservation measures.

It is deplorable that the present British Government appears to react with xenophobic outrage at pressure for observance of some of these international obligations. The EC Special Protection Areas are clearly regarded as an embarrassment, and although the Nature Conservancy Council has now published a detailed list to meet the national requirement for such areas (Stroud *et al.*, 1990), it remains to be seen what response our Government will make. The Government is temporising over a new and helpful EEC measure which would require Member States to undertake formal commitments for habitat safeguard. One hopes that this is a passing phase, and that a maturer wisdom towards international cooperation will eventually prevail. There are signs that the Government recognises the portents of the greenhouse effect and the need for international action to mitigate its impact on humanity. This must be our greatest concern, for without such remedial action, all our other conservation efforts, past and present, for wildlife in this country could be set at nought.

Conclusion

We are concerned here with the future of Nature's last refuge in our islands. Despite the relatively healthy position of our upland bird fauna in general, we cannot afford to be complacent or to depend on good fortune. There is probably a greater number of breeding species than at any time in recorded history, the once endangered Peregrine has made a spectacular recovery, and the struggling Red Kite no longer faces extinction. The reintroduced Sea Eagle is breeding again. Yet many species are on the decline, if only slowly and locally: the Merlin, Hen Harrier (once again), Golden Eagle, Raven, Red Grouse, Golden Plover, Curlew, Snipe, Dunlin, Greenshank and Dipper. The resurgence of montane species appears to be a climatic effect, which could easily be reversed, especially if the greenhouse effect is not reduced by international action.

Even without this wider threat, we must continue to fight for the defence of our mountains and moorlands against continuing damage and loss. To those other interests who may protest against hindrance to further development, and demand to know how much we want, we say 'The same as you: as much as possible'. The conservation strategy should be to minimise all further losses, and only when other parties are prepared to place limits on their ambitions can there be any case for discussing conservation requirements in similar terms. Further development in the uplands matters hardly a fig to the national economy. It is almost entirely a matter of benefit to local, sectional or individual interests, and is mostly heavily dependent on public funds derived largely from the wealth of urban society. Against this must be set the interest of the much larger number of people concerned to protect, cherish and enjoy the national heritage of wild Nature, of which our uplands and their birds are such a vital part.

APPENDIX

Scientific names

Plants

Alder	*Alnus glutinosa*
Ash	*Fraxinus excelsior*
Aspen	*Populus tremula*
Bearberry	*Arctostaphylos uva-ursi*
Bents	*Agrostis* spp.
Bilberry	*Vaccinium myrtillus*
Birch	*Betula pubescens*
	B. verrucosa
Bird cherry	*Prunus padus*
Blackthorn	*P. spinosa*
Bogbean	*Menyanthes trifoliata*
Bogmoss	*Sphagnum* spp.
Bog myrtle	*Myrica gale*
Bracken	*Pteridium aquilinum*
Common gorse	*Ulex europaeus*
Common rush	*Juncus effusus*
Cotton-grass	*Eriophorum vaginatum*
	E. angustifolium
Cowberry	*Vaccinium vitis-idaea*
Crowberry	*Empetrum nigrum*
Deer sedge	*Scirpus caespitosus*
Dwarf azalea	*Loiseleuria procumbens*
Dwarf birch	*Betula nana*
Flying bent (purple moor grass)	*Molinia caerulea*
Great woodrush	*Luzula sylvatica*
Hair moss	*Polytrichum* spp.
Hawthorn	*Crataegus monogyna*
Hazel	*Corylus avellana*
Heather:	
Ling	*Calluna vulgaris*
Bell heather	*Erica cinerea*
Cross-leaved heath	*Erica tetralix*
Holly	*Ilex aquifolium*

237

Juniper	*Juniperus communis*
Larch	*Larix* spp.
Least willow	*Salix herbacea*
Lodgepole pine	*Pinus contorta*
Mat-grass	*Nardus stricta*
Moss campion	*Silene acaulis*
Mossy cyphel	*Cherleria sedoides*
Mountain sedge	*Carex bigelowii*
Northern crowberry	*Empetrum hermaphroditum*
Norway spruce	*Picea abies*
Oak	*Quercus robur, Q. petraea*
Reindeer moss	*Cladonia arbuscula,*
	C. impexa, C. rangiferina
Rose	*Rosa canina* agg.
Rowan	*Sorbus aucuparia*
Saxifrages	*Saxifraga* spp.
Scots pine	*Pinus sylvestris*
Sedges	*Carex* spp.
Sheep's fescue	*Festuca ovina*
Sitka spruce	*Picea sitchensis*
Small-leaved lime	*Tilia cordata*
Spring gentian	*Gentiana verna*
Sycamore	*Acer pseudoplatanus*
Three-pointed rush	*Juncus trifidus*
Thrift	*Armeria maritima*
Vernal grass	*Anthoxanthum odoratum*
Viviparous fescue	*Festuca vivipara*
Wavy hair grass	*Deschampsia flexuosa*
Western gorse	*Ulex gallii*
Willow	*Salix* spp.
Woolly fringe moss	*Racomitrium lanuginosum*
Wych elm	*Ulmus glabra*

Birds

Alpine Accentor	*Prunella collaris*
Alpine Chough	*Pyrrhocorax graculus*
Arctic Redpoll	*Acanthus hornemanni*
Arctic Skua	*Stercorarius parasiticus*
Arctic Tern	*Sterna paradisaea*
Arctic Warbler	*Phylloscopus borealis*
Barn Owl	*Tyto alba*
Bar-tailed Godwit	*Limosa lapponica*
Blackbird	*Turdus merula*
Black Grouse	*Tetrao tetrix*
Black-headed Gull	*Larus ridibundus*
Black-throated Diver	*Gavia arctica*
Blue Rock Thrush	*Monticola solitarius*
Bluethroat	*Cyanosylvia svecica*
Brambling	*Fringilla montifringilla*

Broad-billed Sandpiper	*Limicola falcinellus*
Buzzard	*Buteo buteo*
Capercaillie	*Tetrao urogallus*
Carrion Crow	*Corvus corone*
Chaffinch	*Fringilla coelebs*
Chough	*Pyrrhocorax pyrrhocorax*
Coal Tit	*Parus ater*
Common Gull	*Larus canus*
Common Sandpiper	*Actitis hypoleucos*
Common Scoter	*Melanitta nigra*
Common Tern	*Sterna hirundo*
Crossbill	*Loxia curvirostra*
Cuckoo	*Cuculus canorus*
Curlew	*Numenius arquata*
Curlew Sandpiper	*Calidris ferruginea*
Dartford Warbler	*Sylvia undata*
Dipper	*Cinclus cinclus*
Dotterel	*Charadrius morinellus*
Dunlin	*Calidris alpina*
Dunnock	*Prunella modularis*
Fieldfare	*Turdus pilaris*
Fulmar	*Fulmarus glacialis*
Gannet	*Sula bassana*
Goldcrest	*Regulus regulus*
Golden Eagle	*Aquila chrysaetos*
Goldeneye	*Bucephala clangula*
Golden Plover	*Pluvialis apricaria*
Goosander	*Mergus merganser*
Goshawk	*Accipiter gentilis*
Grasshopper Warbler	*Locustella naevia*
Great Black-backed Gull	*Larus marinus*
Great Northern Diver	*Gavia immer*
Great Skua	*Stercorarius skua*
Great Snipe	*Gallinago media*
Great Tit	*Parus major*
Greenland White-fronted Goose	*Anser albifrons flavirostris*
Green Sandpiper	*Tringa ochropus*
Greenshank	*Tringa nebularia*
Greylag Goose	*Anser anser*
Grey Phalarope	*Phalaropus fulicarius*
Grey Plover	*Pluvialis squatarola*
Grey Wagtail	*Motacilla cinerea*
Guillemot	*Uria aalge*
Gyr Falcon	*Falco rusticolus*
Hen Harrier	*Circus cyaneus*
Heron	*Ardea cinerea*
Herring Gull	*Larus argentatus*
Hooded Crow	*Corvus corone cornix*
House Martin	*Delichon urbica*
House Sparrow	*Passer domesticus*

Jackdaw	*Corvus monedula*
Jack Snipe	*Lymnocryptes minimus*
Jay	*Garrulus glendarius*
Kestrel	*Falco tinnunculus*
King Eider	*Somateria spectabilis*
Kittiwake	*Rissa tridactyla*
Knot	*Calidris canutus*
Lapland Bunting	*Calcarius lapponicus*
Lapwing	*Vanellus vanellus*
Lesser Black-backed Gull	*Larus fuscus*
Lesser Golden Plover	*Pluvialis dominica*
Linnet	*Acanthis cannabina*
Little Stint	*Calidris minuta*
Long-eared Owl	*Asio otus*
Long-tailed Duck	*Clangula hyemalis*
Long-tailed Skua	*Stercorarius longicaudus*
Magpie	*Pica pica*
Mallard	*Anas platyrhynchos*
Manx Shearwater	*Puffinus puffinus*
Marsh Harrier	*Circus aeruginosus*
Meadow Pipit	*Anthus pratensis*
Merlin	*Falco columbarius*
Mistle Thrush	*Turdus viscivorus*
Montagu's Harrier	*Circus pygargus*
Nightjar	*Caprimulgus europaeus*
Osprey	*Pandion haliaetus*
Oystercatcher	*Haematopus ostralegus*
Partridge	*Perdix perdix*
Pectoral Sandpiper	*Calidris melanotos*
Peregrine	*Falco peregrinus*
Pied Flycatcher	*Ficedula hypoleuca*
Pied Wagtail	*Motacilla alba*
Pinkfooted Goose	*Anser brachyrhynchus*
Pintail	*Anas acuta*
Pomarine Skua	*Stercorarius pomarinus*
Ptarmigan	*Lagopus mutus*
Puffin	*Fratercula arctica*
Purple Sandpiper	*Calidris maritima*
Raven	*Corvus corax*
Razorbill	*Alca torda*
Red-backed Shrike	*Lanius collurio*
Red-breasted Merganser	*Mergus serrator*
Red Grouse	*Lagopus lagopus scoticus*
Red Kite	*Milvus milvus*
Red-necked Phalarope	*Phalaropus lobatus*
Redpoll	*Acanthis flammea*
Redshank	*Tringa totanus*
Redstart	*Phoenicurus phoenicurus*
Red-throated Diver	*Gavia stellata*
Red-throated Pipit	*Anthus cervinus*

Redwing	*Turdus iliacus*
Reed Bunting	*Emberiza schoeniclus*
Ring Ouzel	*Turdus torquatus*
Ringed Plover	*Charadrius hiaticula*
Robin	*Erithacus rubecula*
Rock Thrush	*Monticola saxatilis*
Rook	*Corvus frugilegus*
Rough-legged Buzzard	*Buteo lagopus*
Ruff	*Philomachus pugnax*
Sand Martin	*Riparia riparia*
Sanderling	*Calidris alba*
Scaup	*Aythya marila*
Shore Lark	*Eremophila alpestris*
Short-eared Owl	*Asio flammeus*
Siskin	*Carduelis spinus*
Skylark	*Alauda arvensis*
Slavonian Grebe	*Podiceps auritus*
Smew	*Mergus albellus*
Snipe	*Gallinago gallinago*
Snow Bunting	*Plectrophenax nivalis*
Snow Finch	*Montifringilla nivalis*
Snowy Owl	*Nyctea scandiaca*
Song Thrush	*Turdus philomelos*
Sparrowhawk	*Accipiter nisus*
Spotted Redshank	*Tringa erythropus*
Starling	*Sturnus vulgaris*
Stock Dove	*Columba oenas*
Stonechat	*Saxicola torquata*
Swallow	*Hirundo rustica*
Swift	*Apus apus*
Tawny Owl	*Strix aluco*
Teal	*Anas crecca*
Temminck's Stint	*Calidris temminckii*
Tree Pipit	*Anthus trivialis*
Tree Sparrow	*Passer montanus*
Twite	*Acanthis flavirostris*
Turnstone	*Arenaria interpres*
Velvet Scoter	*Melanitta fusca*
Wall Creeper	*Tichodroma muraria*
Water Pipit	*Anthus spinoletta spinoletta*
Waxwing	*Bombycilla garrulus*
Wheatear	*Oenanthe oenanthe*
Whimbrel	*Numenius phaeopus*
Whinchat	*Saxicola rubetra*
White-tailed Eagle	*Haliaeetus albicilla*
Whitethroat	*Sylvia communis*
Whooper Swan	*Cygnus cygnus*
Wigeon	*Anas penelope*
Willow Grouse	*Lagopus lagopus*
Willow Warbler	*Phylloscopus trochilis*
Woodcock	*Scolopax rusticola*

Woodlark	*Lullula arborea*
Wood Pigeon	*Columba palumbus*
Wood Sandpiper	*Tringa glareola*
Wren	*Troglodytes troglodytes*
Yellow Wagtail	*Motacilla flava*
Yellowhammer	*Emberiza citrinella*

Mammals

Bank Vole	*Clethrionomys glareolus*
Bison	*Bison priscus*
Brown Bear	*Ursus arctos*
Brown Hare	*Lepus capensis*
Common Shrew	*Sorex araneus*
Elk	*Alces alces*
Fox	*Vulpes vulpes*
Giant Deer	*Megaloceros giganteus*
Hedgehog	*Erinaceus europaeus*
Horse	*Equus ferus*
Lemmings	*Lemmus, Myopus, Dicrostonyx*
Lion	*Panthera leo*
Mammoth	*Mammuthus primigenius*
Mole	*Talpa europaea*
Mountain Hare	*Lepus timidus*
Musk Ox	*Ovibos moschatus*
Otter	*Lutra lutra*
Pine Marten	*Martes martes*
Rat	*Rattus norvegicus*
Red Deer	*Cervus elaphus*
Reindeer	*Rangifer tarandus*
Short-tailed Field Vole	*Microtus agrestis*
Show-shoe Hare	*Lepus americanus*
Spotted Hyaena	*Crocuta crocuta*
Stoat	*Mustela erminea*
Weasel	*Mustela nivalis*
Wolf	*Canis lupus*
Wood Mouse	*Apodemus sylvaticus*
Woolly Rhinoceros	*Coelodonta antiquitatis*
Yellow-necked Mouse	*Apodemus flavicollis*

Amphibians

Frog	*Rana temporaria*
Toad	*Bufo bufo*

Fish

Brown Trout	*Salmo trutta*
Pike	*Esox lucius*
Salmon	*Salmo salar*

Insects

Antler Moth	*Cerapteryx graminis*
Brackenclock Beetle	*Phyllopertha horticola*
Craneflies	*Tipulidae*
Emperor Moth	*Pavonia pavonia*
Heather Beetle	*Lochmaea suturalis*
Fox Moth	*Macrothylacia rubi*
Northern Eggar Moth	*Lasiocampa quercus callunae*

Other

Sheep Tick	*Ixodes ricinus*

REFERENCES

Adair, P. (1892). The Short-eared Owl and the Kestrel in the vole plague districts. *Annals of Scottish Natural History 1892*, 219–31.

Baines, D. (1988). The effects of improvement of upland, marginal grasslands on the distribution and density of breeding wading birds (Charadriiformes) in northern England. *Biological Conservation* **45**, 221–36.

Bannerman, D.A. (1953–1963). *The Birds of the British Isles*. 12 vols. Edinburgh: Oliver & Boyd.

Barnes, R.F.W. (1987). Long-term declines of Red Grouse in Scotland. *Journal of Applied Ecology* **24**, 735–41.

Bibby, C.J. (1987). Foods of breeding Merlins *Falco columbarius* in Wales. *Bird Study* **34**, 64–70.

Bibby, C.J. & Nattrass, M. (1986). Breeding status of the Merlin in Britain. *British Birds* **79**, 170–85.

Birks, H.J.B. (1988). Long-term ecological change in the British uplands. In *Ecological Change in the Uplands* (ed. M.B. Usher & D.B.A. Thompson), pp. 37–56. Oxford: Blackwell Scientific Publications.

Blezard, E., Garnett, M., Graham, R. & Johnston, T.L. (1943). *The Birds of Lakeland*. Transactions of the Carlisle Natural History Society, Vol. VI.

Bolam, G. (1912). *The Birds of Northumberland and the Eastern Borders*. Alnwick: Blair.

Bolam, G. (1913). *Wildlife in Wales*. London: Frank Palmer.

Booth, C., Cuthbert, M. & Reynolds, P. (1984). *The Birds of Orkney*. The Orkney Press.

Brown, L.H. (1976). *British Birds of Prey*. The New Naturalist. London: Collins.

Brown, L.H. & Watson, A. (1964). The Golden Eagle in relation to its food supply. *Ibis* **106**, 78–100.

Bullock, I.D., Drewett, D.R. & Mickleburgh, S.P. (1983). The Chough in Britain and Ireland. *British Birds* **76**, 377–401.

Bundy, G. (1976). Breeding biology of the Red-throated Diver. *Bird Study* **23**, 249–56.

Bundy, G. (1979). Breeding and feeding observations on Black-throated Divers. *Bird Study* **26**, 33–6.

Cadbury, C.J. (1987). Moorland birds – Britain's international responsibility. *RSPB Conservation Review*, No. 1, 1987. Sandy: Royal Society for the Protection of Birds.

Cadbury, C.J., Elliott, G. & Harbard, P. (1988). Birds of prey conservation in the UK. *RSPB Conservation Review*, No. 2, 1988. Sandy: Royal Society for the Protection of Birds.

Cade, T.J., Enderson, J.H., Thelander, C.G. & White, C.M. (eds) (1988). *Peregrine*

Falcon Populations: their Management and Recovery. Boise, Idaho: The Peregrine Fund, Inc.

Campbell, B., Watson, A. Snr, Watson, A. & Picozzi, N. (1974). Proof of breeding of Shore Larks. *British Birds* **67**, 127.

Chapman, A. (1907). *Bird-life of the Borders*. 2nd edition. London: Gurney & Jackson.

Chapman, A. (1924). *The Borders and Beyond*. London: Gurney & Jackson.

Condry, W.M. (1981). *The Natural History of Wales*. The New Naturalist. London: Collins.

Coulson, J.C. (1956). Mortality and egg production of the Meadow Pipit, with special reference to altitude. *Bird Study* **3**, 119–32.

Cramp, S. & Simmons, K.E.L. (1977–88). *Handbook of the Birds of Europe, the Middle East and North Africa. The Birds of the Western Palaearctic*. Vols I–V. Oxford University Press.

Cumming, I.G. (1979). Lapland Bunting breeding in Scotland. *British Birds* **72**, 53–6.

Dare, P.J. (1986a). Raven *Corvus corax* populations in two upland regions of north Wales. *Bird Study* **33**, 179–89.

Dare, P.J. (1986b). Aspects of the breeding biology of Ravens in two upland regions of North Wales. *Naturalist* **111**, 129–37.

Davies, M. (1988). The importance of Britain's Twites. *RSPB Conservation Review*, No. 2, 1988. Sandy: Royal Society for the Protection of Birds.

Davies, P.W. & Davis, P.E. (1973). The ecology and conservation of the Red Kite in Wales. *British Birds* **66**, 183–224; 241–70.

Davis, P.E. & Davis, J.E. (1981). The food of the Red Kite in Wales. *Bird Study* **28**, 33–40.

Davis, P.E. & Davis, J.E. (1986). The breeding biology of a Raven population in central Wales. *Nature in Wales* **3**, 44–54.

Davis, P.E. & Newton, I. (1981). Population and breeding of Red Kites in Wales over a 30 year period. *Journal of Animal Ecology* **50**, 759–72.

Dennis, R.H. (1983). Purple Sandpipers breeding in Scotland. *British Birds* **76**, 563–6.

Dennis, R.H., Ellis, P.M., Broad, R.A. & Langslow, D.R. (1984). The status of the Golden Eagle in Britain. *British Birds* **77**, 592–607.

Eriksson, M.O.G. (1986). Reproduction of Black-throated Diver *Gavia arctica* in relation to fish density in oligotrophic lakes in south-western Sweden. *Ornis Scandinavica* **17**, 245–8.

Evans, A.H. (1911). *A Fauna of the Tweed Area*. Edinburgh: David Douglas.

Fisher, J. (1949). *Natural History of the Kite*. RSPB Occasional Publication No. 8. London: Royal Society for the Protection of Birds.

Fryer, G. (1986). Notes on the breeding biology of the Buzzard. *British Birds* **79**, 18–28.

Furness, R.W. (1983). *The Birds of Foula*. The Brathay Hall Trust.

Furness, R.W. (1987). *The Skuas*. Calton: Poyser.

Gladstone, H.S. (1930). *Record Bags and Shooting Records*. London: Witherby.

Glutz von Blotzheim, U.N., Bauer, K.M. & Bezzel, E. (1975) *Handbuch der Vögel Mitteleuropas*, vol. 6. Wiesbaden: Akademische Verlagsgesellschaft.

Godwin, H. (1975). *History of the British Flora*. Cambridge University Press.

Gordon, S. (1955). *The Golden Eagle, King of Birds*. London, Collins.

Graham, R. (1934). Bird notes from the Solway, Pennines and Lakeland, 1932, 1933. *North Western Naturalist* **9**, 135–49; 345–56.

Green, R.E. (1986). *Breeding waders of the Somerset Moors: factors affecting their distribution and breeding success*. Report to the Nature Conservancy Council, Contract No. HF3/03/291.

Greenhalgh, M.E. (1974). The Pennine gullery. *Bird Study* **21**, 146–8.

Harrison, C.J.O. (1987). Pleistocene and prehistoric birds of south-west Britain. *Proceedings of the University of Bristol Spelaeological Society* **18**, 81–104.

Harvie-Brown, J.A. (1906). *A Fauna of the Tay Basin and Strathmore*. Edinburgh: David Douglas.

Harvie-Brown, J.A. & Buckley, T.E. (1895). *A Fauna of the Moray Basin*. 2 vols. Edinburgh: David Douglas.

Harvie-Brown, J.A. & Buckley, T.E. (1887). *A Vertebrate Fauna of Sutherland, Caithness and West Cromarty*. Edinburgh: David Douglas.

Harvie-Brown, J.A. & Macpherson, H.A. (1904). *A Fauna of the North-West Highlands and Skye*. Edinburgh: David Douglas.

Haviland, M.D. (1926). *Forest, Steppe and Tundra*. Cambridge University Press.

Hewson, R. & Leitch, A.F. (1982). The spacing and density of Hooded Crow nests in Argyll (Strathclyde). *Bird Study* **29**, 235–8.

Hickey, J.J. (ed.) (1969). *Peregrine Falcon Populations: their Biology and Decline*. Madison: University of Wisconsin Press.

Holdsworth, M. (1971). Breeding biology of Buzzards at Sedbergh during 1937–67. *British Birds* **64**, 412–20.

Holmes, R.T. (1970). Differences in population density, territoriality and food supply of Dunlin on arctic and sub-arctic tundra. In *Animal Populations in Relation to their Food Resources* (ed. A. Watson), pp. 303–19. Oxford: Blackwell.

Holyoak, D. (1971). Movements and mortality of Corvidae. *Bird Study* **18**, 97–106.

Hudson, P. (1986*a*). *Red Grouse. The Biology and Management of a Wild Gamebird*. Fordingbridge: The Game Conservancy Trust.

Hudson, P. (1986*b*). The effect of a parasitic nematode on the breeding production of Red Grouse. *Journal of Animal Ecology* **55**, 85–92.

Hudson, P. & Renton, J. (1988). What influences changes in grouse numbers? Game Conservancy Report for 1987. Fordingbridge: The Game Conservancy.

Hudson, P. & Watson, A. (1985). Exploited animals. The Red Grouse. *The Biologist* **32**, 13–18.

Jenkins, D., Watson, A. & Miller, G.R. (1963). Population studies on Red Grouse *Lagopus lagopus scoticus* (Lath.) in north-east Scotland. *Journal of Animal Ecology* **32**, 317–76.

Lack, D. (1954). *The Natural Regulation of Animal Numbers*. Oxford University Press.

Lack, P. (ed.) (1986). *The Atlas of Wintering Birds in Britain and Ireland*. Calton: Poyser.

Lamb, H.H. (1982). *Climate, History and the Modern World*. London: Methuen.

Leitch, A.F. (1986). Report on eagle predation on lambs in the Glenelg area in 1986. Peterborough: Nature Conservancy Council.

Leslie, A.S. (ed.) (1912). *The Grouse in health and in disease*. Popular edition. London: Smith, Elder & Co.

Lockie, J.D. (1955*a*). The breeding and feeding of Jackdaws and Rooks, with notes on Carrion Crows and other Corvidae. *Ibis* **97**, 341–69.

Lockie, J.D. (1955*b*). The breeding habits of Short-eared Owls after a vole plague. *Bird Study* **2**, 53–69.

Lockie, J.D. (1964). The breeding density of the Golden Eagle and Fox in relation to food supply in Wester Ross, Scotland. *The Scottish Naturalist* **71**, 67–77.

Lockie, J.D. & Ratcliffe, D.A. (1964). Insecticides and Scottish Golden Eagles. *British Birds* **57**, 89–102.

Lockie, J.D., Ratcliffe, D.A. & Balharry, R. (1969). Breeding success and organochlorine residues in Golden Eagles in west Scotland. *Journal of Applied Ecology* **6**, 381–9.

Love, J.A. (1988). The re-introduction of the White-tailed Sea Eagle to Scotland:

1975–1987. *Research and survey in nature conservation*, no. 12. Peterborough: Nature Conservancy Council.

MacArthur, R.H. & Wilson, E.O. (1967). *The Theory of Island Biogeography*. Princeton University Press.

Mackenzie, O. (1924). *A Hundred Years in the Highlands*. London: Arnold.

Macpherson, H.A. (1892). *A Vertebrate Fauna of Lakeland*. Edinburgh: David Douglas.

Manley, G. (1952). *Climate and the British Scene*. The New Naturalist. London: Collins.

Marquiss, M. & Newton, I. (1982). The Goshawk in Britain. *British Birds* **75**, 243–60.

Marquiss, M., Newton, I. & Ratcliffe, D.A. (1978). The decline of the Raven *Corvus corax* in relation to afforestation in southern Scotland and northern England. *Journal of Applied Ecology* **15**, 129–44.

Marquiss, M., Ratcliffe, D.A. & Roxburgh, R. (1985). Breeding success and diet of Golden Eagles in southern Scotland in relation to change in land use. *Biological Conservation* **34**, 121–40.

Massey, M.E. (1978). The bird community of an upland nature reserve in Powys, 1970–1977. *Bird Study* **25**, 167–73.

McVean, D.N. & Lockie, J.D. (1969). *Ecology and land use in upland Scotland*. Edinburgh University Press.

Mearns, R. (1983). The status of the Raven in southern Scotland and Northumbria. *Scottish Birds* **12**, 211–18.

Mills, D.H. (1962). The Goosander and Red-breasted Merganser as predators of salmon in Scottish waters. *Freshwater and Salmon Fisheries Research*, no. 29.

Milsom, T.P. & Watson, A. (1984). Numbers and spacing of summering Snow Buntings and snow cover in the Cairngorms. *Scottish Birds* **13**, 19–23.

Moore, N.W. (1957). The past and present status of the Buzzard in the British Isles. *British Birds* **50**, 173–97.

Moss, R. (1969). A comparison of Red Grouse *(Lagopus l. scoticus)* stocks with the production and nutritive value of heather *(Calluna vulgaris)*. *Journal of Animal Ecology* **38**, 103–12.

Moss, R. & Watson, A. (1980). Inherent changes in the aggressive behaviour of a fluctuating red grouse *Lagopus lagopus scoticus* population. *Ardea* **68**, 113–19.

Moss, R. & Watson, A. (1985). Adaptive value of spacing behaviour in population cycles of Red Grouse and other animals. In *Behavioural Ecology: Ecological Consequences of Adaptive Behaviour* (ed. R.M. Sibley & R.H. Smith), pp. 275–94. Oxford: Blackwell Scientific Publications.

Nature Conservancy Council (1984). *Nature conservation in Great Britain*. London: Nature Conservancy Council.

Nature Conservancy Council (1986). *Nature conservation and afforestation in Britain*. Peterborough: Nature Conservancy Council.

Nelson, T.H. (1907). *The Birds of Yorkshire*. 2 vols. London: A. Brown.

Nethersole-Thompson, D. (1951). *The Greenshank*. The New Naturalist. London: Collins.

Nethersole-Thompson, D. (1966). *The Snow Bunting*. Edinburgh: Oliver & Boyd.

Nethersole-Thompson, D. (1971). *Highland Birds*. Inverness: Highlands and Islands Development Board.

Nethersole-Thompson, D. (1973). *The Dotterel*. London: Collins.

Nethersole-Thompson, D. (1976). Recent distribution, ecology and breeding of Snow Buntings in Scotland. *Scottish Birds* **9**, 147–62.

Nethersole-Thompson, D. & Nethersole-Thompson, M. (1979). *Greenshanks*. Berkhamsted: Poyser.

Nethersole-Thompson, D. & Nethersole-Thompson, M. (1986). *Waders, their Breed-*

ing Haunts and Watchers. Calton: Poyser.

Nethersole-Thompson, D. & Watson, A. (1981). *The Cairngorms*, 2nd edition. Perth: The Melven Press.

Newton, I. (1979). *Population Ecology of Raptors*. Berkhamsted: Poyser.

Newton, I., Davis, P.E. & Davis, J.E. (1982). Ravens and Buzzards in relation to sheep farming and forestry in Wales. *Journal of Applied Ecology* **19**, 681–706.

Newton, I. & Haas, M.B. (1988). Pollutants in Merlin eggs and their effects on breeding. *British Birds* **81**, 258–69.

Newton, I., Meek, E.R. & Little, B. (1978). Breeding ecology of the Merlin in Northumberland. *British Birds* **71**, 376–98.

Newton, I., Meek, E.R. & Little, B. (1984). Breeding season foods of Merlin *Falco columbarius* in Northumberland. *Bird Study* **31**, 49–56.

Newton, I., Meek, E & Little, B. (1986). Population and breeding of Northumbrian Merlins. *British Birds* **79**, 155–70.

Newton, I., Robson, J.E. & Yalden, D.W. (1981). Decline of the Merlin in the Peak District. *Bird Study* **28**, 225–34.

Ogilvy, R.S.D. (1986). Whither forestry? The scene in AD 2025. In *Trees and Wildlife in the Scottish Uplands* (ed. D. Jenkins), pp. 33–9. ITE Symposium No 17. Huntingdon: Institute of Terrestrial Ecology.

Orford, N. (1973). Breeding distribution of the Twite in central Britain. *Bird Study* **20**, 51–62, 121–6.

Ormerod, S.J., Allinson, N., Hudson, D. & Tyler, S.J. (1986). The distribution of breeding Dippers (*Cinclus cinclus* (L.) Aves) in relation to stream acidity in upland Wales. *Freshwater Biology* **16**, 501–7.

Ormerod, S.J., Tyler, S.J. & Lewis, J.M.S. (1985). Is the breeding distribution of Dippers influenced by stream acidity? *Bird Study* **32**, 33–9.

Parr, R. (1979). Sequential breeding by Golden Plovers. *British Birds* **72**, 499–503.

Parr, R. (1980). Population study of Golden Plover *Pluvialis apricaria*, using marked birds. *Ornis Scandinavica* **11**, 179–89.

Pearsall, W.H. (1950). *Mountains and Moorlands*. The New Naturalist. London: Collins.

Petty, S.J. & Anderson, D. (1986). Breeding by Hen Harriers *Circus cyaneus* on restocked sites in upland forests. *Bird Study* **33**, 177–8.

Philipson, M. (1954). North-eastern bird studies. The Curlew and the Black Grouse. In *Lakeland Ornithology* (ed. E. Blezard), pp. 14–22. *Transactions of the Carlisle Natural History Society*, vol. 8.

Phillips, J.S. (1970). Inter-specific competition in Stonechat and Whinchat. *Bird Study* **17**, 320–4.

Picozzi, N. (1971). Breeding performance and shooting bags of Red Grouse in relation to public access in the Peak District National Park, England. *Biological Conservation* **3**, 211–15.

Picozzi, N. (1978). Dispersion, breeding and prey of the Hen Harrier, *Circus cyaneus*, in Glen Dye, Kincardineshire. *Ibis* **120**, 498–509.

Picozzi, N. (1984). Sex ratio, survival and territorial behaviour in polygynous Hen Harriers *Circus c. cyaneus* in Orkney. *Ibis* **126**, 356–65.

Picozzi, N. & Weir, D. (1976). Dispersal and causes of death in Buzzards. *British Birds* **69**, 193–201.

Potts, G.R., Tapper, S.C. & Hudson, P.J. (1984). Population fluctuations in Red Grouse: analysis of bag records and a simulation model. *Journal of Animal Ecology* **53**, 21–6.

Poxton, I.R. (1986). Breeding Ring Ouzels in the Pentland Hills. *Scottish Birds* **14**, 44–8.

Poxton, I.R. (1987). Breeding status of the Ring Ouzel in south-east Scotland. *Scottish Birds* **14**, 205–8.

Rankin, N. (1947). *Haunts of British Divers*. London: Collins.

Ratcliffe, D.A. (1962). Breeding density in the Peregrine *Falco peregrinus*, and Raven *Corvus corax*. *Ibis* **104**, 13–39.

Ratcliffe, D.A. (1976). Observations on the breeding of the Golden Plover in Great Britain. *Bird Study* **23**, 63–116.

Ratcliffe, D.A. (ed.) (1977). *A Nature Conservation Review*. 2 vols. Cambridge University Press.

Ratcliffe, D.A. (1980). *The Peregrine Falcon*. Calton: Poyser.

Ratcliffe, D.A. (1986). The effects of afforestation on the wildlife of open habitats. In *Trees and Wildlife in the Scottish Uplands* (ed. D. Jenkins), pp. 96–54. ITE Symposium No. 17. Huntingdon: Institute of Terrestrial Ecology.

Ratcliffe, D.A. & Thompson, D.B.A. (1988). The British uplands: their ecological character and international significance. In *Ecological Change in the Uplands* (ed. M.B. Usher & D.B.A. Thompson), pp. 9–36. Oxford: Blackwell Scientific Publications.

Reed, T. (1981). The number of breeding landbird species on British islands. *Journal of Animal Ecology* **50**, 613–24.

Richardson, M.G. (1990). The distribution and status of Whimbrel (*Numenius p. phaeopus*) in Scotland and Britain. *Bird Study*, **37**, 61–8.

Roberts, J.L. & Bowman, N. (1986). Diet and ecology of Short-eared Owls *Asio flammeus*, breeding on heather moor. *Bird Study* **33**, 12–17.

Robson, R.W. (1956). The breeding of the Dipper in north Westmorland. *Bird Study* **3**, 170–80.

Seebohm, H. (1883). *A History of British Birds*. Vol. 1. London: Porter & Dulan.

Seel, D.C. & Walton, K.C. (1979). Numbers of Meadow Pipits *Anthus pratensis* on mountain farm grassland in north Wales in the breeding season. *Ibis* **121**, 147–64.

Sharrock, J.T.R. (1976). *The Atlas of Breeding Birds in Britain and Ireland*. Tring: British Trust for Ornithology.

Shaw, G. (1978). The breeding biology of the Dipper. *Bird Study* **25**, 149–60.

Shooter, P. (1970). The Dipper population of Derbyshire, 1958–68. *British Birds* **63**, 158–63.

Snow, D.W. (1968). Movement and mortality among British Kestrels (*Falco tinnuculus*). *Bird Study* **15**, 65–83.

Stroud, D.A., Reed, T.M., Pienkowski, M.W. & Lindsay, R.A. (1987). *Birds, Bogs and Forestry. The Peatlands of Caithness and Sutherland*. Peterborough: Nature Conservancy Council.

Stroud, D.A., Condie, M., Holloway, S.J., Rothwell, A.J., Shepherd, K.B., Simons, J.R. & Turner, J. (1988). *A survey of moorland birds on the Isle of Lewis in 1987*. Internal CSD Report No. 776. Peterborough: Nature Conservancy Council.

Stroud, D.A., Mudge, G.P. & Pienkowski, M.W. (1990). *Protecting internationally important bird sites. A review of the EEC Special Protection Area network in Great Britain*. Peterborough: Nature Conservancy Council.

Taylor, K., Hudson, R. & Horne, G. (1988). Buzzard breeding distribution and abundance in Britain and northern Ireland in 1983. *Bird Study* **35**, 109–18.

Thom, V. (1986). *Birds in Scotland*. Calton: Poyser.

Thomas, C.J. & Tasker, M.L. (1988). *The Abbeystead Gull Colony in 1988*. Contract report No 844. Peterborough: Nature Conservancy Council.

Thompson, D.B.A. & Thompson, M.L.P. (1985). Early warning and mixed species association: the Plover's Page revisited. *Ibis* **127**, 559–62.

Thompson, D.B.A., Thompson, P.S. & Nethersole-Thompson, D. (1986). Timing of breeding and breeding performance in a population of Greenshanks. *Journal of Animal Ecology* **55**, 181–99.

Thompson, D.B.A., Stroud, D.A. & Pienkowski, M.W. (1988). Afforestation and upland birds: consequences for population ecology. In *Ecological Change in the Uplands* (ed. M.B. Usher & D.B.A. Thompson), pp. 237–59. Oxford. Blackwell Scientific Publications.

Tyler, S. (1987). River birds and acid water. In *RSPB Conservation Review 1987* (ed. C.J. Cadbury & M. Everett), pp. 68–70. Sandy: Royal Society for the Protection of Birds.

Usher, M.B. & Thompson, D.B.A. (1988). *Ecological Change in the Uplands*. (British Ecological Society special publication no. 7). Oxford: Blackwell Scientific.

Ussher, R.J. & Warren, R. (1900). *The Birds of Ireland*. London: Gurney & Jackson.

Village, A. (1982). The home range and density of Kestrels in relation to vole abundance. *Journal of Animal Ecology* **51**, 413–28.

Voous, K.H. (1960). *Atlas of European Birds*. Amsterdam: Nelson.

Wanless, S. & Langslow, D.R. (1983). The effects of culling on the Abbeystead and Mallowdale gullery. *Bird Study* **30**, 17–23.

Watson, A. (1963). Sections on Ptarmigan in the northern Highlands, under Scottish Ptarmigan. In *The Birds of the British Isles*, Vol XII, by D.A. Bannerman. Edinburgh: Oliver & Boyd.

Watson, A. (1965). A population study of Ptarmigan (*Lagopus mutus*) in Scotland. *Journal of Animal Ecology* **34**, 135–72.

Watson, A. & Moss, R. (1980). Advances in our understanding of the population dynamics of Red Grouse from a recent fluctuation in numbers. *Ardea* **68**, 103–11.

Watson, A. & Moss, R. (1987). Scottish Grouse. *Shooting Times*. October 29–November 4, pp. 22–3.

Watson, A. & Rae, R. (1987). Dotterel numbers, habitats and breeding success in Scotland. *Scottish Birds* **14**, 191–8.

Watson, A., Payne, S. & Rae, R. (1989). Golden Eagles *Aquila chrysaetos*: land use and food in northeast Scotland. *Ibis* **131**, 336–348.

Watson, D. (1977). *The Hen Harrier*. Berkhamsted: Poyser.

Watson, J., Langslow, D.R. & Rae, S.R. (1987). *The impact of land-use changes on Golden Eagles in the Scottish Highlands*. Contract Report No 720. Peterborough: The Nature Conservancy Council.

Weir, D. & Picozzi, N. (1983). Dispersion of Buzzards on Speyside. *British Birds* **76**, 66–78.

Williamson, K. (1951). The moorland birds of Unst, Shetland. *Scottish Naturalist* **63**, 37–44.

Williamson, K. (1975). Birds and climatic change. *Bird Study* **22**, 143–64.

Witherby, H.F., Jourdain, F.C.R., Ticehurst, N.F. & Tucker, B.W. (1938). *The Handbook of British Birds* (5 vols). London: Witherby.

Woods, A. & Cadbury, C.J. (1987). Too many sheep in the hills. In *RSPB Conservation Review 1987* (ed. C.J. Cadbury & M. Everett), pp. 65–7. Sandy: Royal Society for the Protection of Birds.

Yalden, D.W. (1986). The habitat and activity of Common Sandpipers *Actitis hypoleucos*, breeding by upland rivers. *Bird Study* **33**, 214–22.

Yalden, D.W. & Yalden, P.E. (1988). *Golden Plovers and Recreational Disturbances*. Internal 3rd Report to NCC on contract no. HF 3-03-339. Peterborough: Nature Conservancy Council.

Yeates, G.K. (1948). *Bird haunts in northern Britain*. London: Faber & Faber.

INDEX

Only the more important bird and place names are included, and the information of Table 1 is not indexed. Principal references are printed in bold.